# TRANSACTIONAL TO TRANSFORMATIONAL MARKETING IN PHARMA

*The Science of Why and The Art of How*

# TRANSACTIONAL TO TRANSFORMATIONAL MARKETING IN PHARMA

*The Science of Why and The Art of How*

**Subba Rao Chaganti**

**PharmaMed Press**
*An imprint of BSP Books Pvt. Ltd.*
4-4-309/316, Giriraj Lane,
Sultan Bazar, Hyderabad - 500 095.

**Transactional to Transformational Marketing in Pharma**
**The Science of *Why* and The Art of *How***
**by *Subba Rao Chaganti***



***Published by***

**PharmaMed Press**
*An imprint of BSP Books Pvt. Ltd.*
4-4-309/316, Giriraj Lane, Sultan Bazar, Hyderabad - 500 095.
Phone: 040-23445688, 23445600; Fax: 91+40-23445611
E-mail: info@pharmamedpress.com
www.pharmamedpress.com/pharmamedpress.net

**Cover Design**

**J. Siva Prasad, Amitha Designs**

**ISBN: 978-81-961469-0-0 (Hardbound)**

**Dedicated to:**

Sri J. M. Raju and Srimathi Rukmini for being a constant source of inspiration, encouragement, and support since publishing my first book.

# ACKNOWLEDGEMENTS

First and foremost, I must acknowledge my appreciation and thanks to Andrew Witty, the CEO of GlaxoSmithKline, who put the company on a revolutionary and transformational route to marketing pharmaceuticals, setting the pace for the rest of the industry in 2013.

Andrew Witty took charge as the CEO of GlaxoSmithKline 2008, a top-ten global pharmaceutical company. Later in 2013, he faced a crisis with scandals such as the China bribery. As a result, the company had to pay a fine of $490 million in China. In addition, the company had to pay one of the largest settlement fees—about $3 billion to the US Department of Justice, for many unethical practices.

In 2013, Andrew Witty launched a new ethical approach to its interactions with customers, transforming the company's marketing model to overcome the negative perceptions and image of the company.

Andrew Witty's transformational marketing model and the GSK team that showed great courage and conviction in carrying it through at the time are the main sources of inspiration for me to write this book— *Transactional to Transformational Marketing in Pharma: The Science of Why and The Art of How*.

While GlaxoSmithKline has set the pace for transformational marketing for the pharma majors globally, a small-to-medium-size company, Curatio, operating in a very small niche segment of derma-cosmetics in the fiercely competitive Indian pharmaceutical market, has also followed a transformational marketing route when most of the industry is quagmired in transformational marketing. P. V. Shankar Dass, CEO, and G. K. Ramani, Founder-Director of Curatio, showed great courage and conviction in building the company by following a transformational marketing route to a ₹ 2,200 crores value at which they recently sold the company to Torrent, a leading international generics company from India. Shankar Dass and Ramani give me abundant confidence that even a small and medium-sized pharmaceutical company can practice transformational marketing, leveraging technology. One doesn't necessarily have very deep pockets to invest in technology to transform marketing processes.

However, one can prioritize investing in technology in a phased manner, defining milestones.

A few individuals' help and support has been significant in bringing out this book, *Transactional to Transformational Marketing: The Science of Why and The Art of How*. Without their physical, intellectual, and moral support, this book would still be in the conceptual stage. I sincerely thank:

- Mr. J. Sivaprasad, managing director of Amitha Designs, for designing a fantastic cover for the book
- Anil Shah is the most enthusiastic publisher with a passion for publishing that I have known
- Nikunjesh Shah for making the digital versions of this book available
- Mr. Naresh Daver and his team and PharmaMedPress for designing this book beautifully

Last but not least, I owe my thanks to Mahalakshmi, my better half, both literally and figuratively, for supporting me throughout the writing of this book. I must also thank my three children and their spouses—Srinivasa Phanindra, Geetha, Lavanya, Aditya, Soumya, and Chaitanya for their appreciation and feedback during this book project. In addition, my son, Srinivasa Phanindra, has been very encouraging with his constructive feedback on the initial draft of this book.

More importantly, I thank my grandchildren— Aditi, Eesha, Surya, and Shriya. They have been a constant source of joy, happiness, and inspiration that contributed greatly to the writing of this book. The sparkle in their eyes when they saw their names as I wrote them in acknowledgments gave me immense satisfaction.

**Subba Rao Chaganti**

# PROLOGUE

Despite its great contribution to improving the health and well-being of people and saving countless lives worldwide for over a century, the modern pharmaceutical industry has continued to fall from grace in public perception for the last ten to fifteen years.

While there are many reasons, such as ever-escalating prices denying access to many patients, unethical business practices to gain market share, and relentless focus on the bottom line explain for low productivity, they cannot be responsible for Pharma's loss of reputation. Instead, most pharmaceutical companies' transactional marketing and other unethical business practices seem to be a major, if not the only, cause of the drug industry's fall on the reputation barometer.

Increasing competition, drying up of new product pipelines, rapid genericization, and continuing cost-containment pressures are what a marketer practicing transactional marketing would offer as possible reasons. But, then, there is often a very thin line between a reason and an excuse. Excuses, when accepted, become reasons, and reasons, when denied, become excuses.

Regardless of whether there are reasons or excuses, *Transactional Marketing* practices and unethical business behavior may give short-term gains in business at a huge cost to reputation, which is extremely difficult to rebuild. Moreover, transactional relationships are so transient, like fleeting clouds, and they cannot build loyalty for your brand or company. Only *Transformational Marketing* practices and ethical business behavior can.

Pharma must transform its marketing practices from transactional to transformational today. The question is not whether pharma can, but how?

The book, *Transactional to Transformational Marketing in Pharma* aims to show how a pharma company can do this because the cure for all current marketing ills in Pharma is Transformational Marketing!

**Subba Rao Chaganti**

# CONTENTS

## CHAPTER 3



## CHAPTER 4



## CHAPTER 5

# CHAPTER 6

## Transformational Marketing in Pharma 119-284

# LIST OF ABBREVIATIONS

| | |
|---|---|
| AARP | American Association of Retired Persons |
| APPROVe | Adenomatous Polyp Prevention on Vioxx |
| ADHD | Attention Deficit Hyperactivity Disease |
| API | Application Protocol Interface |
| AI | Artificial Intelligence |
| AR | Augmented Reality |
| ARDSI | Alzheimer's and Related Disorders Society of India |
| AWS | Amazon Web Services |
| BMS | Bristol-Myers Squibb |
| BPaaS | Business Process as a Service |
| CAGR | Compound Annual Growth Rate |
| CETIFARMA | Council of Ethics and Transparency of the Pharmaceutical Industry |
| CIO | Chief Information Officer |
| CLM | Closed Loop Marketing |
| CME | Continuing Medical Education |
| COPD | Chronic Obstructive Pulmonary Disease |
| CSR | Corporate Social Responsibility |
| CTA | Call To Action |
| CTR | Click Through Rate |
| DOJ | Department of Justice |
| DOP | Department of Pharmaceuticals |
| DPI | Dry Powder Inhaler |
| DSCSA | Drug Supply Chain Security Act |
| DSMB | Data and Safety Monitoring Board |
| DTCA | Direct-To-Customer Advertising |
| EDC | Electronic Data Capture |
| EFPIA | European Federation of Pharmaceutical Industries and Associations |
| ePRO | Electronic Patient Reported Outcomes |

| | |
|---|---|
| ERP | Enterprise Resource Planning |
| FCPA | Foreign Corrupt Practices Act |
| FCRA | Foreign Contribution Regulation Act |
| FDA | Federal Drug Administration |
| FMRAI | Federation of Medical and Sales Representatives Association of India |
| GAO | General Accounting Office |
| GSK | GlaxoSmithKline |
| GSMA | Global Systems for Mobile Communications Association |
| HCP | Health Care Professional |
| HIPAA | Health Insurance Portability and Accountability Act |
| HRT | Hormonal Replacement Therapy |
| IDMA | Indian Drug Manufacturers Association |
| IFPMA | International Federation of Pharmaceutical Manufacturers & Associations |
| IND | Investigational New Drug |
| KOL | Key Opinion Leader |
| KPI | Key Performance Indicator |
| MCM | Multi Channel Marketing |
| MDI | Metered Dose Inhaler |
| MS | Multiple Sclerosis |
| MSL | Medical Science Liaison |
| MLR | Medical, Legal, Regulatory |
| NEJM | New England Journal of Medicine |
| NPS | Net Promoter Score |
| NSAID | Non-Steroidal Anti Inflammatory Drug |
| OA | Osteoarthritis |
| OPPI | Organization of Pharmaceutical Producers of India |
| OTC | Over-the-Counter |
| PaaS | Platform as a Service |

| | |
|---|---|
| PDA | Personal Digital Assistant |
| PhRMA | Pharmaceutical Research and Manufacturers of America |
| PIR | Physician Information Request |
| PPC | Pay-Per-Click |
| PSR | Professional Service Representative |
| RDPAC | R&D-based Pharmaceutical Association Committee |
| RDS | Remote Desktop Services |
| ROI | Return on Investment |
| SEA | Search Engine Advertising |
| SEC | Securities Exchange Commission |
| SEM | Search Engine Marketing |
| SEO | Search Engine Optimization |
| SERP | Search Engine Result Pages |
| SFA | Sales Force Automation |
| SMS | Short Messaging Service |
| SNS | Simple Notification Services |
| UCPMP | Uniform Code of Pharmaceutical Marketing Practices |
| UNCAC | United Nations Convention Against Corruption |
| URL | Uniform Resource Locator |
| VAS | Value Added Service |
| VIGOR | Vioxx Gastrointestinal Outcomes Research |
| VR | Virtual Reality |

CHAPTER

1

# Pharma's Reputation on A Slide

About a quarter of the money taken by pharmaceutical companies for the drugs they sell is turned around into promotional activity which has, as we will see , a provable impact on doctors' prescribing. So, we pay for products, with huge uplift in price to cover their marketing budgets, and that money is then spent on distorting evidence-based practice, which in turn makes our decisions unnecessarily expensive, and less effective.

—Ben Goldcare, Author of *Bad Pharma: How Drug Companies Mislead Doctors, and Harm Patients*

# Pharma's Reputation on A Slide

## Declining Reputation

Pharma's reputation has been going down these days. Pharma bashing, too, has become a popular game. The public perception of the pharmaceutical industry is currently at the lowest it has been in recent history. Consider, for example, the case of Merck, the company which was Fortune Magazine's most admired company in the US for an unprecedented seven years in a row. Paradoxically, it was the same Merck, the marketer of Vioxx, a product that experienced one of the most well-publicized drug recalls and ultimate withdrawal. Several books, such as *Bad Pharma* by Ben Goldacre, *The truth about the drug companies* by Marcia Angell, and *Hooked* by Howard Brody, described several areas of controversy, such as unethical marketing practices and lack of transparency. Magazines such as Forbes have devoted stories calling the industry 'Pill Pushers' and detailing how pharmaceutical companies have 'abandoned science for salesmanship.'

## Good Practices

The behavior and practices of pharmaceutical companies determine whether their reputation is going north or south. In other words, what you do determines your reputation. Not very long ago, the pharmaceutical industry enjoyed a great reputation and even the admiration of all the stakeholders and the general public. They conquered many fatal diseases because of their significant contribution to society through their breakthrough discoveries in medicine. Here are some of the more important good practices that Pharma engages in :

- The industry plays a major role in discovering and developing new medicines.
- Responsible education of healthcare professionals (in particular physicians)
- The understanding of the true value of drugs for the appropriate patient populations. The recent advances towards

personalized healthcare by many specialty pharmaceutical companies indicate this.

- The Patients' quality of life improved with better diagnosis and compliance.
- The continuous use of clinical research supports the true value addition of drugs.
- Building up evidence concerning the needs and wants of patients to the R&D community and allocating more resources to the appropriate research projects.
- Planning and implementing patient-centric strategies from bench to bedside.
- Pharma companies have established R&D centers to work on cures for neglected diseases. Pharmaceutical companies are devoting resources to finding treatment for malaria, trypanosomiasis (sleeping sickness), dengue fever, and Chagas disease that plague the developing world. Many companies are working on these projects with the Gates Foundation, the United Nations Children's Fund (UNICEF), and World Health Organization (WHO).
- Companies such as GlaxoSmithKline have been running the African Malaria Partnership for a decade to implement behavioral change programs to aid malaria prevention in vulnerable areas.
- The industry has been known for its philanthropy too. In annual surveys of the most generous companies, pharmaceutical companies dominate the list.

## Pharma's Bad Practices

Pharmaceutical companies have received their share of criticism in recent years, not only concerning their alleged profiteering but also for their behavior that led to excessive profits. The pharmaceutical industry's marketing practices have been particularly facing severe criticism. The areas which face criticism more often are:

- Excessive incentives for company sales reps
- Using their medical liaisons to promote products

- Excessive incentives to Key opinion leaders (KOLs) and physicians for prescribing the drugs
- Physician engagement practices. Campbell et al. in 2007 wrote in an article, *A national survey of physician-industry relationships* published in the New England Journal of Medicine, that out of 3,167 physicians surveyed, 94 percent of physicians had free food in their offices; 28 percent received consultancy fees for lectures or clinical trial recruiting; 35 percent received reimbursement for attending continuing medical education programs (CMEs) or professional meetings. This is only illustrative of the nature of physician engagement with Pharma.
- Off-label promotions
- Lack of transparency and deception over outcomes and scientific evidence for marketed products.
- The pharmaceutical industry spends more money on marketing than on research and development, and this marketing expenditure drives drug prices very high. In 2014, Global Data reported that Johnson & Johnson, Pfizer, and Novartis were spending almost double their R&D expenses on marketing while others, such as Roche and Lilly, almost equal amounts on marketing and R&D.
- Bribery: Transparency International in 2011 reported that the pharmaceutical industry ranked seventh out of 19 industries that use bribery to speed up administrative processes in 30 countries worldwide. GlaxoSmithKline's bribery charges in China, and Kickbacks to pharmacies in the US by Novartis are widely known.

These unethical marketing practices have been responsible for bringing the pharmaceutical industry's reputation down and not the marketing of prescription drugs per se. Lea Prevel Katsanis, a marketing professor, in her insightful book *Global Issues in Pharmaceutical Marketing*, suggested the following factors associated with reputational damage:

1. The pharmaceutical industry is an industry that is more inward-looking and resistant to change, with self-reinforcing beliefs about its marketing practices.

2. In-experienced marketing managers who manage brands and learn their skills on the job without sufficient formal training.
3. A lack of accuracy and balance in the presentation of marketing messages. The effects of this message multiply as it is repeated in multiple channels.
4. The use of direct-to-consumer advertising (DTCA) and how it trivializes drug therapy.
5. The belief is that if a particular marketing activity is legal, then by definition, it must also be appropriate. Pharmaceutical companies have a 'no apologies' approach regarding their marketing activities.
6. The perception of high drug prices is a consequence of sizable marketing budgets. The public, in particular, believes that pharmaceutical companies spend more on marketing than on research to develop new drugs, and this marketing activity results in higher prices.
7. The effects of physician engagement with the industry result in a potential bias toward prescribing the drugs.
8. The inaccurate medical news reporting and the blurred lines between legitimate news and marketing messages.
9. The perception is that the pharmaceutical industry sets its own agenda to determine disease treatment policies.
10. The public distrust of the industry results from the negative publicity given to off-label drug marketing.

## Pharma's Bad Practices: Cases

Here are a few cases that illustrate some bad marketing practices by leading pharmaceutical companies providing the pharmaceutical marketer with valuable insights into what not to do and avoid.

CASE

1

# The Vioxx Fiasco!

Merck discovered Vioxx (Rofecoxib), a Cox 2 selective inhibitor, by a team led by P. Prasit at Merck-Frosst in Montreal, Canada. Merck acquired Charles E. Frosst in 1965. FDA approved Vioxx in May 1999.

Vioxx was an anti-inflammatory drug used to treat arthritis and acute pain without stomach irritation caused by other non-steroidal anti-inflammatory (NSAID) drugs. Merck promoted Vioxx aggressively using direct marketing to doctors, private clinics, and hospitals through advertising campaigns both in print media and television. The drug was also endorsed by celebrities who were former athletes, such as Dorothy Hamill and Bruce Jenner. Merck also offered Vioxx to hospitals and doctors at discounted rates. As a result, Vioxx emerged as one of the best-selling drugs in treating arthritis and acute pain within one year of launch.

Termed Super Aspirin, Vioxx was promoted as a cure for everything from arthritis pain to menstrual cramps. In addition, it was projected as a pain reliever, which was a boon for patients suffering from arthritis. Moreover, Vioxx relieved pain without gastrointestinal problems caused by older-generation painkillers. Merck spent about $160.8 million promoting Vioxx in 1999. It soon emerged as one of the fastest-selling drugs in the world. Vioxx quickly became a blockbuster drug for treating the pain associated with osteoarthritis (OA) and rheumatoid arthritis (RA).

However, even as the prescriptions and sales were increasing rapidly for Vioxx, so were the concerns about its side effects. Although Vioxx was considered gastro-safe as there were no gastrointestinal side effects, the same thing cannot be said about cardiovascular side effects. Medical experts have been raising doubts about the cardiovascular risks associated with Vioxx's long-term usage almost since its launch. In the initial years, Merck disagreed with the various medical studies that indicated cardiovascular risks until its own internal study indicated the risk.

Merck's VIGOR clinical studies published in 2004, showed a five-fold increase in myocardial infarction among patients taking Vioxx compared to patients taking Naproxen. In September 2004, clinical trials showed that Vioxx increased the risk of myocardial infarction and stroke. Immediately, Merck voluntarily withdrew Vioxx from the market on September 30.

The stock market reacted violently to Merck's withdrawal of Vioxx from the market, wiping out about $28 billion of the company's stock value in a few months.

Launched in May 1999, and withdrawn on September 30, 2004. During the short life of Vioxx, it had about 105 million prescriptions in the US. More than 84 million people had taken the drug worldwide, and at the time of recall, about 2 million were taking it. Soon after the recall, Merck's share prices fell by 27 percent from $45.07 to $33 per share, wiping out $28 billion in market value. Vioxx had been the fastest-moving drug in Merck's portfolio at the time of recall. What a fall! How did it happen? The timeline of events leading to the precipitous fall of Vioxx is presented in the following table.

**Table 1.1 The Rise and Fall of Vioxx: A Timeline**

| Period | Actions and Activities |
|---|---|
| November 1998 | Merck completes a clinical trial testing their Investigative New Drug (IND) on 5,400 subjects and seeks US FDA approval. |
| January 1999 | Merck launches Vioxx Gastrointestinal Outcomes Research study (VIGOR) with more than 8,000 participants. Half of the participants take Vioxx and the other half take Naproxen, a pain killer and NSAID of an older generation. The clinical trial is designed to see whether Vioxx is safer for the gastrointestinal system than Naproxen. |
| May 1999 | FDA approves Vioxx, making the drug available by prescription in the US. |
| October 1999 | The VIGOR Study's Data and Safety Monitoring Board (DSMB) meets for the first time and notes that Vioxx patients have fewer ulcers and less gastrointestinal bleeding than patients taking Naproxen. It looks as if the study will be a success for Merck. |
| November 1999 | The VIGOR Study's Data and Safety Monitoring Board (DSMB) meets again and focuses on heart problems. The panel finds that 79 patients out of 4,000 taking Vioxx have serious heart problems or have died, compared with 41 patients taking Naproxen. They note that while the trends are disconcerting, the number of events are small, and to continue the study. |
| December 1999 | The safety panel observes that the risk of serious heart problems and death among Vioxx patients are twice as high as in the Naproxen group. The panel, while recommending the continuation of the study, decides that Merck needs to develop a plan to analyze the study's cardiovascular results before the study ends.<br>When recommending the continuation of the study, the Safety panel said that it could not tell if Vioxx was causing the heart problems or if Naproxen, acting like a low-dose aspirin protected people from them, making Vioxx just look risky in comparison. |
| January 2000 | Merck hesitates at developing the analysis plan. The company wants to wait and combine the cardiovascular results of VIGOR with the results from other Vioxx studies. But, Dr. Michael E. Weinblatt, the safety panel Chair and a rheumatologist from Brigham & Women's Hospital at Boston pushes for immediate analysis.<br>Merck agrees to analyze heart problems reported reported by February 10 - at least a month before the last patient leaves the study. Events reported later won't be included in the initial analysis. |

**Table 1.1** *Contd...*

| Period | Actions and Activities |
|---|---|
| February 2000 | Dr. Michel E. Weinblatt files out a disclosure form that says he and his wife own $ 72,975 of Merck stock. He also agrees to a new contract involving 12 days of work over two years, at the rate of $ 5,000 per day. |
| March 2000 | Merck gets the results of the VIGOR trial. |
| May 2000 | Merck submits VIGOR study paper to the New England Journal of Medicine (NEJM) for publication. The data includes only 17 out of 20 heart attacks Vioxx patients had. |
| July - November 2000 | A. A memo from Merck statistician Deborah Shapiro to Merck scientist Alise Reicin (both are listed authors of the NEJM paper) refers to heart attacks 18, 19, and 20 suffered by patients taking Vioxx during the study.<br>B. Merck tells the FDA about the heart attacks 18, 19, and 20.<br>C. VIGOR study authors submit two sets of corrections to their NEJM manuscript, without mentioning the three additional heart attacks<br>D. NEJM publishes the VIGOR study results, still with no mention of the three additional heart attacks in the Vioxx group. The published results also leave out many other kinds of cardiovascular adverse events. |
| February 2001 | The FDA holds advisory meeting on VIGOR trials. It publishes complete VIGOR data on its website, including the additional heart attacks and data on other cardiovascular events. |
| August, 2001 | Cardiologists - Debabrata Mukherjee, Steven Wissen, and Eric Topol publish their meta-analysis in the Journal of American Medical Association (JAMA), based on complete VIGOR data that the FDA has made available. |
| January 2002 to August 2004 | Numerous epidemiological studies point out that Vioxx increased risk of cardiovascular problems. |
| September 2004 | A. Merck withdraws Vioxx after its colon-polyp prevention study -APPROVe shows that the drug raises the risk of heart attacks after 18 months. By the time Vioxx is withdrawn from the market, an estimated 20 million Americans have taken the drug.<br>B. The decision resulted in a huge loss of $ 28 billion in market value for Merck.<br>C. The number of lawsuits blaming Vioxx for the deaths of patients who were taking the drug also start mounting. |

**Table 1.1** *Contd...*

| Period | Actions and Activities |
|---|---|
| July 2005 | NEJM editor-in-chief, Dr. Jeffrey M. Drazen tells NPR that the journal had been hoodwinked by Merck and that the authors of the VIGOR study should have told the Journal about the additional heart attacks. |
| November / December 2005 | A. NEJM issues an *Expression of Concern* writing that, inaccuracies and deletions in the VIGOR manuscript Merck submitted to the Journal call into question the *integrity of data*. The Journal asks the study authors to submit a correction to the Journal.<br>B. The first Federal trial on Vioxx lawsuits begins in 2005 in New Jersey court. |
| March 2006 | VIGOR study authors respond to NEJM's expression of concern stating: our evaluation leads us to conclude that our original article follows appropriate clinical trial principles and does not require a correction. The three additional heart attacks in question occurred after the study's respecified cutoff date for reporting cardiovascular problems. |
| May 2006 | A. Outside analysis of data sent to the FDA from Vioxx APPROVe study show that the cardiovascular risks from Vioxx began shortly after patients started taking the drug. The data also indicated that the risks from Vioxx remain long after patients stop taking the drug.<br>B. Merck disagrees with the analysis and maintains that patients are not at risk unless they had taken the drug for more than 18 months. This point is very important for Merck as it could cost the company billions of dollars. Many of those suing the company say that they took the drug for less than 18 months. |
| June 2006 | A. The seventh trial against Merck begins. Out of six cases that have already gone to trial, Merck has won three and lost three.<br>B. Research published in the medical journal, Lancet estimates that 88,000 Americans had heart attacks from taking Vioxx and 38,000 of them died. |
| November 2007 | Merck announces that it will pay $4.85 billion to end thousands of lawsuits over its painkiller Vioxx. The amount paid into a settlement fund is believed to be the largest settlement ever. |

Source: Snigdha Prakash, Vikki Valentine, The Rise and Fall of Vioxx: A Timeline, NPR, November 10, 2007

Vioxx's withdrawal had cost Merck dearly in loss of revenues, market capitalization, and reputation. What is paradoxical about the whole Vioxx fiasco is that Merck was banking on a gastro-safe drug and compared it with a drug, Naproxen, that is cardiac-safe but has serious gastrointestinal side effects like many NSAIDs. In the bargain, it has a gastro-safe drug with serious and fatal cardiovascular side effects. As Dr. Wayne Ray, an Epidemiologist at Vanderbilt University, aptly observed, "*A heart attack in exchange for an ulcer is a poor treatment.*"

(Source: Snigdha Prakash, Vikki Valentine, The Rise and Fall of Vioxx: A Timeline, NPR, November 10, 2007)

CASE

2

# Marketing through Manipulation and Misinformation!

Parke-Davis, one of the leading American pharmaceutical firms, patented Gabapentin (Neurontin) in 1977, and obtained US FDA approval in 1993, as adjunctive therapy for partial-complex seizures. Neurontin became a significant blockbuster for Parke-Davis, which later became a division of Warner-Lambert by an acquisition. In 2000, Pfizer acquired Warner-Lambert.

Sales of Neurontin rose from US $95 million in 1995 to nearly $3 billion in 2004, after which its patent expired, and it lost most of its sales to the generics that flooded the market. How Neurontin became a three-billion dollar molecule is a story that is stranger than fiction and tells us that success at any cost is very costly! Here is a brief account of what happened:

## The Early 1990s

Parke-Davis was facing serious challenges in the early 1990s. The patents of the company's blockbuster drugs had expired. The R&D pipeline, too, was nothing to write home about. Moreover, the stock market had downgraded the firm's stock. The company was in desperate need of something good to sell. The company hired a marketing consultant, Richard Vanderveer, with considerable knowledge of pharmaceutical marketing research and strategy. He helped the company institute a micro-marketing program.

### Micro-marketing Program

Micro-marketing program is about targeting individual physicians with tailored information that resonates with them as individuals. The communication is target-specific or physician-specific, considering individual physicians' needs. It, therefore, would be highly interesting to doctors, unlike a carpet-bombing approach, where the same data and information are presented to all the physicians.

Around the same time, the company also hired Anthony Wild, who had considerable experience in the sales and marketing side of the pharmaceutical industry. He knew the task at hand when he joined Parke-Davis and was well aware of the crucial nature of his assignment. Moreover, he had a compelling need to succeed.

### The 'Wild' Era

The company was banking all its hopes for future survival and growth on the two new drugs pending approval with the US FDA. One was Lipitor (Atorvastatin), a cholesterol-reducing drug; the other was Rezulin (Troglitazone) for treating type 2 diabetes.

Wild chalked out a survival plan for himself as well as the company. First, he identified three products among the existing product mix of Parke-Davis, which were not selling well, but had great potential. These products were - Neurontin (approved as an adjunct therapy in epilepsy), Accupril (Quinapril), an anti-hypertensive drug, and Loestrin, a low-estrogen birth control pill. He planned to raise the sales of each of these three brands to reach a 15 percent market share in their respective therapeutic categories. Moreover, he wanted to invest the profits generated by these three products in promoting the two new drugs once the company received approval from the FDA. He proposed his plan to the top management and got their approval.

Wild had set out to change the culture at the company. He focused on the following:

1. Change the somewhat fatalistic attitude at the company to a high-confident one. His message to the sales force? Believe in yourself!
2. From a risk-averse or low-risk stance to a high-risk, rich-rewards mindset
3. Removing all the caps or restrictions on sales force incentives and making them very attractive
4. Relentless focus on increasing the sales of Neurontin

### Focus on Expanding Usage of Neurontin

Although the primary approval for Neurontin was in the adjunctive therapy of partial-complex seizures, the sales force was hearing favorable comments from doctors who had experimented with off-label uses of the drug to treat neuropathic pain, bipolar disorders, attention deficit disorders, migraine, restless legs syndrome, alcohol, and drug withdrawal and as a mono-therapy for seizures (instead of an adjuvant). Yet no reliable evidence proved Neurontin's benefits in treating these conditions. It was all anecdotal.

Conducting clinical trials for new drug applications was the only way to establish the efficacy of Neurontin in all these conditions. However, clinical trials were very expensive and were not guaranteed success. Moreover, Neurontin was coming off patent at four year-end; therefore, the huge investment in clinical trials would not be viable. Although charged with the new incentive system and all revved up, the sales force seemed helpless in expanding the sales. How could they increase sales without overtly promoting its off-label use, which was illegal? The company found its answer in the medical liaisons division. The primary responsibility of a medical liaison is to provide fair and balanced scientific information regarding clinical trials, drug uses, side effects, and adverse reactions and help physicians understand the state of the science and up-to-date information on the treatment modalities. They should not engage in any way in persuading physicians to prescribe their products. The medical liaison executives are usually medical doctors or PhDs and should have the domain expertise comparable to physicians they visit to maintain a peer-to-peer status.

The company knew promoting its products and soliciting prescriptions through its medical liaison team was illegal, but it continued. But unfortunately, the company committed several unethical and even illegal actions in the process. Here is a very brief account of their so-called innovative marketing practices and what they did:

1. The company started hiring medical liaisons directly out of the sales department. They were all trained in sales techniques to generate prescriptions that the company needed very badly.
2. They also started incentivizing the medical liaisons by partly compensating them based on sales.
3. The medical liaisons had to work with the regular pharmaceutical representatives of the company and had no communication with the medical research division. The company gave their medical science liaisons a list of doctors for 'cold calls' based on the size of the doctors' practices and their potential to prescribe Neurontin. In addition, the company provided them with a package of monetary incentives to offer physicians for participating in the Parke-Davis programs.
4. Although it is illegal for a drug company to pay physicians to prescribe a drug, paying them to be special consultants is not technically illegal. Parke-Davis paid thousands of physicians to become such consultants. It is not a coincidence that all the physicians who received money from the company had one thing in common. They were all heavy prescribers of Neurontin, particularly for treating neuropathic pain.
5. At the time, Parke-Davis implemented a Preceptor program in which physicians who allowed a company representative to

visit included discussions with patients. As a part of the Preceptor program, representatives often had an opportunity to meet actual patients and influence the physicians often to prescribe Neurontin for off-label uses to treat these patients.

6. In addition, the company established a Parke-Davis speaker's bureau and paid high-prescribers of Neurontin *thought leaders* to go out and spread the word. However, there was no substantial clinical data. They only had data from their less rigorous case studies based on their clinical experience where they used Neurontin. The physicians were paid for these case studies on a case-by-case basis to write up each patient's history and response to Neurontin.

All these marketing practices led to a blatant off-label promotion of Neurontin. In April 1996, John Ford, the Parke-Davis Executive, articulated the company's expectations at a recorded marketing managers meeting. He said:

*I want you out there every day selling Neurontin. Look, this isn't just me. It's come down from Morris Plains (Headquarters) that Neurontin is more profitable than Accupril. So we need to focus on Neurontin. Neurontin is not growing for adjunctive therapy (The approved indication). Besides, that's not where the money is. Pain management, now that's the money. Monotherapy, that's the money. We don't want to share these patients with everybody. We want them on Neurontin only. The whole thing is the drug budget, not a quarter or half. We can't wait for them to ask. We need to get out there and tell them out front. Dinner programs, CME (Continuing Medical Education) programs, and consultantships work great but don't forget the one-on-one. That's where we need to be, holding their hand and whispering in the ear, Neurontin for pain, Neurontin for mono-therapy, Neurontin for bipolar, Neurontin for everything. I don't want to see a patient coming off Neurontin before they've been unto 4,800 milligrams daily. I don't want to hear that safety crap either—you should take one just to see if there is nothing. It's a great drug.*

The scale and magnitude of this change in Toni Wild's Parke-Davis were stunning and illegal. While the change apparently energized the marketing team significantly, the legal downside too was substantial. As was mandatory, in April 1996, the company conducted a program to educate medical liaisons about the prevailing legal environment governing their role. A former FDA official and a company lawyer held a seminar on the subject. The seminar was in two parts; only the first part was videotaped. The FDA official and the lawyer said to the team:

*If caught violating the FDA rules, you're on your own and acting without the company's knowledge or permission. You must have a physician information request (PIR) for each call. You must provide a fair and balanced presentation. You cannot close or sell. You can't promote a drug off-label. You cannot promote a drug pre-approval. You must keep an accurate record of your activities. You cannot solicit an inquiry.*

After this, the video camera was turned off, and the second part of the seminar began. The second part of the presentation was candid and to the point. First, the former FDA official gave them tips on circumventing the system without getting caught. Then, it told the team what the company expected from them explicitly:

*We expect you to do your job and stay focused on sales. Don't worry about this (the first part of the seminar). If you are cold calling a sales representative, have him fill out a physician information request form to cover you. The doctors know that you're not out there to help the competitors. So don't worry about being balanced in your presentation. Look, without sales, there is no Parke-Davis. We all have to sell at the same level. Be careful about this. Just don't leave anything behind. Above all, don't put anything in writing.*

The medical liaisons soon became an integrated part of the sales and marketing department. The company gave them the same pep talk and offered the same incentives as it did for the sales teams. The company promised them an all-expense-paid cruise to the Bahamas if it achieved the sales goals for Neurontin and Accupril. A marketing executive told them:

*The only way we will make it (to the Bahamas) is if you as a group take ownership of the task and get out there and aggressively move market share. You have to be aggressive. Don't take no for an answer. If the rep does not close, you close. If the rep sees the wrong doctors, you see the right ones. If a high-prescribing practice is not using Neurontin, get in there, do your thing, then ask why. I don't care, but you're wasting your time and our money if you don't ask for the new prescription when you are through.*

Medical science liaisons soon dominated the Neurontin team. They were contacting more and more high-prescribing physicians as per the micro-marketing strategy. As a result, the physicians prescribed Neurontin for treating many off-label uses, such as bipolar disorder, neuropathic pain, and others. In addition, they focused on incentivizing these doctors and engaging them as part of the Neurontin family. As a result of all these activities, Neurontin achieved explosive growth reaching from a modest US $98 million in 1995 to a whopping US $3 billion in annual sales in 2004.

Finally, all the company's unethical and illegal promotional activities in promoting Neurontin had come to light because of a whistleblower, a new recruit in medical science liaisons. David Franklin, a post-doctoral fellow in microbiology from Harvard University, joined the medical sales liaisons at Parke-Davis on April 1, 1996. He attended the now infamous seminar that the former FDA official and the company lawyer gave the new trainees in medical science liaisons on April 16, 1996. A senior marketing executive repeatedly told Franklin to go out and sell Neurontin for off-label uses, contrary to the briefing he received at the training time. Franklin was disenchanted, disappointed, and confused by these conflicting messages. They were in total contradiction to what he perceived the medical science liaison job would be and the way it had turned out. He left Parke-Davis within three months, collected data and evidence about the firm's illegal marketing practices promoting Neurontin, and filed a suit against the company. Later, Pfizer acquired Warner and Lambert; thus, Parke-Davis became a part of Pfizer in 2000.

After protracted hearings and a detailed investigation into the allegations filed by Franklin, Pfizer pleaded guilty to its illegal marketing practices and agreed to pay $430 million to resolve all the criminal and civil liabilities. The following table presents a timeline describing Neurontin's rise and the company's fall concerning its unethical marketing practices.

**Table 1.2 Illegal Marketing of Neurontin: A Timeline**

| Period | Event |
|---|---|
| 1977 | Parke-Davis obtains a patent for its Gabapentin. |
| 1993 | US FDA approves Gabapentin under the brand name Neurontin as adjunctive therapy for partial-complex seizures. |
| 1995 | Neurontin records US $ 95 million in annual sales. |
| April 1, 1996 | David Franklin, a post-doctoral fellow in microbiology from Harvard University joins Parke-Davis in Medical Science Liaison (MSL). |
| April 16, 1996 | David Franklin and his peers get a briefing from Parke-Davis on FDA regulations and about their responsibilities as medical sceince liaison officers. |
| April 22, 1996 | A senior marketing executive a week later repeatedly tells David Franklin to go and promote the drug Neurontin for off-label uses, which is contrary to the briefing Franklin had at the time of training. |
| August 1996 | Franklin leaves Parke-Davis and files a suit against the company stating that it is indulging in illegal marketing practices such as off-label promotion of its drug Neurontin and making false claims to elicit payments from the Federal government. Case put under seal, deferring action pending government review. |
| December 1999 | The government lifts the seal and litigation resumes. |
| 2000 | Pfizer acquires Warner Lambert along with its Parke-Davis division. |
| October 2000 | The FDA approves Gabapentin (Neurontin) for adjunctive treatment of partial-seizures in children of 3 - 12 years. |
| May 2002 | The FDA approves Gabapentin (Neurontin) for post-herpetic neuralgia in adults. |
| 2004 | Annual sales of Neurontin reach almost US $ 3 billion. |
| May 13, 2004 | Warner-Lambert pleads guilty and agrees to pay US $ 430 million to resolve criminal charges and civil liabilities. |

(Adapted from Greg Critser's book, Generation Rx: How Prescription Drugs Are Altering American Lives, Minds, and Bodies, Houghton Mifflin and Company, New York, 2005)

## Predatory Pricing

Another bad marketing practice that pharmaceutical companies indulge is predatory pricing. Predatory pricing is the illegal activity of setting prices to eliminate competition and create a monopoly. Predatory pricing is also called undercutting. Usually, firms practicing predatory pricing strategies keep their prices low to make them unattractive to competitors. However, a new type of predatory pricing seems to be rearing its ugly head. It is predatory pricing in a reverse direction. Here the pricing strategy followed is to keep the prices very high, even to socially unacceptable levels, and yet keep competition at bay. When you can price your product so high, it must attract more competition, right? That is a logical question. But then, the modus operandi or the business model of these modern pharmaceutical predators on the prowl beats the logic. Their business model looks somewhat like this:

1. **Source a Sole-Source Drug**: Acquire a drug that had come off patent and yet remained a single source drug, which means that there is no imminent competition.
2. **Ensure that it is a Gold Standard**: Check that the single-source drug you acquired or are about to acquire is considered a gold standard for the condition it treats. If it is a gold-standard treatment, the physicians will continue to prescribe it even if the price increases. The perceived efficacy standard and the essential nature of the drug determine the level of price increase that it can absorb.
3. **Select a Drug that has a Smaller Market:** A smaller market means that it is relatively unattractive for competitors to enter. But ensure that the drug has smaller, dependent patient populations, who are too small to organize an opposition to price hikes.
4. **Closed Distribution**: Make sure when you acquire the drug meets all these criteria that it has a closed distribution system and is not easily available from any other sources. When the drug is unavailable through the normal channels, it creates another entry barrier for competition.

Once you acquire a drug meeting all the criteria mentioned above, the firm is ready to practice its predatory pricing strategy, which is in the reverse direction. While predatory pricing is pricing the product or service, keeping it so unattractively low that it becomes an entry barrier to competition. The new predatory pricing is to create entry barriers into their carefully chosen markets and then hike prices exorbitantly to maximize profitability.

CASE

3

# Daraprim's Fifty-fold Price Increase!

Nobel Prize-winning American scientist Gertrude Elion developed Pyrimethamine (Brand Name: Daraprim) at Burroughs-Wellcome (GlaxoSmithKline, now) in 1952 to combat malaria. Pyrimethamine is on the World Health Organization's list of Essential Medicines because it is useful in treating parasitic infections such as Toxoplasmosis and Malaria and is often used for people with compromised immune systems, including AIDS and some forms of cancer, and elderly patients. Later in 2010, GlaxoSmithKline sold the marketing rights for Daraprim to CorePharma, which Impex laboratories later acquired in March 2015. Impex Laboratories sold the rights of Daraprim for the US market to Turing Pharmaceuticals in August 2015.

After purchasing Daraprim, Turing Pharmaceuticals increased the price of a single tablet almost fifty-fold from $13.50 to $750, raising the annual cost of treatment for some patients to hundreds of thousands of dollars. The price increase sparked widespread criticism. For example, the Infectious Diseases Society of America and the HIV Medicine Association sent a joint letter to Turing, stating the increase to be unjustifiable for the medically vulnerable patient population and unsustainable for the healthcare system.

In defense of Daraprim's price increase, the hedge-fund manager turned founder of Turing Pharmaceuticals, Martin Shkreli, said that many patients use Daraprim for less than a year, and the price is more in line with other drugs for rare diseases. Moreover, Daraprim is 0.01 percent of healthcare costs in the US. Further, he promised to negotiate volume discounts for hospitals. He also claimed that a tablet would only cost $1for patients without insurance. Finally, he claimed: I'm like Robin Hood... I'm taking Walmart's money and researching diseases no one cares about, and the money from profits would be used to develop new and better drugs.

Several experts opined that price increase was unjustifiable on any count. For example, Dr. Wendy Armstrong, professor of infectious diseases at Emory University, said in response: *An old drug is not necessarily bad. Daraprim) happens to be an incredibly effective drug and has been cheap and well-tolerated by patients for years*.

Dr. Judith Alberg of Icahn School of Medicine at Mount Sinai said that *Daraprim would be too expensive for hospitals to keep in stock, and the use of the drug would require special review, possibly forcing the hospitals to seek alternative therapies that may not have the safe efficacy. Daraprim's price increase seems to be all profit-driven for somebody. I think it's a very dangerous process*.

Max Nisen wrote in Bloomberg and The Washington Post that: *Old medicines are sold at inflated prices because there's no mechanism to compel drug makers to lower them. So instead, pharmaceutical companies justify drug prices by reminding the public that developing drugs is costly and failure-prone. That's a fair point. But drug companies also announced more than $50 billion worth of share buybacks and dividend hikes after the new 2017 tax-cut law passed*.

In September 2017, Turing Pharmaceuticals became Vyera Pharmaceuticals and started marketing Daraprim in the US under the new company, Vyera Pharmaceuticals. The company responded to the 2015 criticisms of Daraprim with various patient affordability and access initiatives and reduced the cost of Daraprim to hospitals by up to 50 percent.

In January 2020, the FTC filed a case against Vyera alleging an elaborate anticompetitive scheme to preserve a monopoly for the life-saving drug Daraprim. In December 2021, Vyera Laboratories reached a settlement where the company agreed to provide up to $40 million in relief over ten years, to consumers who allegedly were fleeced by their actions and required to make Daraprim available to any potential generic competitor at the cost of producing the drug.

The Daraprim price increase did cause a lot of public outrage, and the CEO of Turing Pharmaceuticals, Martin Shkreli, was put behind bars for wire fraud. However, despite the outrage and legal action, prices for Daraprim have not decreased— one tablet still costs $750 at the time of writing in December 2022.

The concern here is that some predator marketers like Martin Shkreli scouring for drugs like Daraprim that don't have active generic rivals (as the market for such drugs is too small for a generic drug company to view it as profitable). So it is not a question of merely acquiring a drug and raising its price after acquiring the drug. Rather the issue is that these drug companies are strategically searching for drugs that can sustain massive price increases. It is a nefarious motivation, and there lies the moral fault.

(Source: Adapted from articles—(1) Ethics Unwrapped, Daraprim Price Hike, https://ethicsunwrapped.utexas.edu/video/daraprim-price-hike, (2) Ethics Unwrapped, Daraprim Price Hike, https://ethicsunwrapped.utexas.edu/video/daraprim-price-hike, (3) Wandtv.com, Prosecutors: Company that Illegally Monopolized Life-Saving Drug Must Pay $40 million, https://www.wandtv.com/news/prosecutors-company-that-illegally-monopolized-life-saving-drug-must-pay-40m/article_f16bea1a-5794-11ec-918c-af8ad2428046.html, and (4) Vyera Pharmaceuticals. (2022, August 13). In *Wikipedia*. https://en.wikipedia.org/wiki/Vyera_Pharmaceuticals, (5) Fritz Alhoff, *Daraprim and Predatory Pricing: Martin Shkreli's 5000% Hike* on Law and Biosciences Blog, Stanford Law Schools (SLS), Blogs, Stanford Law School)

CASE

4

# Awareness Campaigns, Lobbying, Legislation, Competitors' Stumbles and Exorbitant Price Hikes Make A Blockbuster!

Can you believe that a company that did not know whether to keep a product that it acquired took it in-house, built a multi-pronged marketing strategy to market, and made it a blockbuster, a Go-to product for patients with severe allergies? The company is Mylan, and the product is their auto-injector device, EpiPen. Here is how it all happened.

Epinephrine (also known as adrenaline) that the body produces to increase blood flow to the muscles in its response to fight or flight. Jockichi Takamine, a Japanese chemist, was among the first to discover and isolate epinephrine. Soon after the discovery, scientists figured out how to produce it in large enough quantities and how to use it in different medical settings. In 1906, scientists synthesized epinephrine for the first time. Doctors continued investigating how adrenaline worked and have used it for over a hundred years. It has been studied extensively, with over 12,000 studies referencing it. It has heralded many areas of emergency medication. Epinephrine is used in hospitals worldwide and is on the World Health Organization's (WHO) essential medicines list. It only costs a few dollars a vial in the developed world and much less in the developing world.

In the 1970s, biochemical engineer Sheldon Kaplan invented a way to self-inject epinephrine called ComboPen. Initially, the US military used the ComboPen to protect their soldiers in the event of chemical warfare, as it was easy to use in an emergency situation. Shortly after, Kaplan and others found out that they could use ComboPen to deliver epinephrine in emergencies to treat severe allergic reactions without the presence and help of healthcare providers. The US FDA approved the EpiPen as we knew it now in 1987. Meridian Medical Technologies, now a subsidiary of Pfizer, owned the product.

Later, Merck KGaA, a German drug company, acquired the product. Finally, in 2007, Mylan bought the generics business of Merck KGaA and became the owner of the EpiPen. Interestingly, Meridian continues to manufacture EpiPen for Mylan even today. In 2007, when Mylan acquired EpiPen, its annual sales were around US $200 million. In 2015, EpiPen's global sales passed the coveted one-billion-dollar mark. So how did Mylan achieve this? Here are some of the significant strategic steps that the company took:

1. Mylan specialty, the marketer and distributor of EpiPen auto-injector, launched many allergy and anaphylaxis awareness campaigns, both unbranded and branded, with celebrities living with or caring for someone with severe allergy conditions. The company ran, more recently, a *Face Your Risk* campaign, an ultra-realistic commercial about someone having an allergic reaction to peanut butter. Mylan spent a billion dollars to raise awareness of the need for EpiPens in the eight years from 2009 to 2016.
2. Mylan had also invested in a huge lobbying effort and got legislation passed in 48 states allowing schools to have undesignated EpiPens for emergency use.
3. Mylan has increased the price of a pack of two EpiPen auto-injectors exorbitantly from $93 in 2008 to $608.61 in 2016.

Mylan today has a virtual monopoly of the epinephrine market in the US, with over 90 percent market share. While these are the three main reasons for the phenomenal sales growth of Epipen, its price increase has been the most controversial and has drawn criticism from all corners of society. In response, the company increased its copay coupon system and doubled the discount to $300 on a two-pack EpiPen auto-injector. In addition, the company announced sometime back that it would introduce a cheaper version of the EpiPen at half the current price.

The phenomenal sales success is due to many factors, such as awareness campaigns, lobbying leading to legislative changes, competitors' inability to field an approvable alternative, and an exorbitant price hike of 500 percent— some acceptable and some unacceptable.

### Lessons

Mylan followed the classic marketing strategy behind every blockbuster drug meticulously. These are impactful disease awareness campaigns and lobbying to get the legislation to empower schools to provide EpiPen auto-injectors in time to treat medical emergencies due to severe allergic reactions.

Perhaps the most important lesson is that one should not exploit their monopoly situation and price it irresponsibly just to maximize profits at the expense of patients.

(Source: Adapted from articles— (1) Emily Willingham, Why Did Hike EpiPen Prices 400%? Because They Could, forbes.com, August 21, 2016, (2) Sarah Kliff, EpiPEn's 400 percent Price Hike Tells Us A Lot About What's Wrong With American Healthcare, vox.com, August 23, 2016)

CHAPTER

2

# Ethics in the Pharmaceutical Industry

Ethics are knowing the difference between
What you have a right to do
And
What is right to do!

*— Potter Stewart, Former Associate Justice of the United States Supreme Court*

## Ethics in the Pharmaceutical Industry

What are ethics in the pharmaceutical industry? Ethics in the pharmaceutical industry are the moral principles and values that govern drug research, development, and marketing.

The pharmaceutical industry plays a crucial role in contributing to the health and well-being of society by discovering, developing, and producing new treatments for diseases. However, it faces ethical challenges in commercializing the treatments it develops, such as conflicts of interest, pricing, and access to medicines. For example, consider the following ethical challenges that the industry faces:

- One of the major ethical issues in the pharmaceutical industry is the relationship between drug companies and healthcare providers. For example, pharmaceutical companies often provide financial incentives, such as research grants, consulting fees, and speaker fees, to key opinion leaders for speaking about their products at medical symposia and continuing medical education programs, which can influence the prescribing of drugs, leading to conflicts of interest.
- Another ethical issue is the high cost of drugs, which can limit access to necessary treatments for many patients, particularly those uninsured or underinsured.
- DTCA (Direct-to-Consumer-Advertising) is another issue that has been attracting criticism. Critics argue that DTCA can lead to over-diagnosis, over-treatment, and higher costs. However, DTCA for prescription drugs is not approved in most countries worldwide except in the US and New Zealand.

Thus, ethics in the pharmaceutical industry is a complex and multifaceted issue that encompasses a range of moral principles and values, including transparency, fairness, and access to medicine. Pharma companies must collaborate with various stakeholders, such as healthcare providers, policymakers, regulatory authorities, and patient advocacy groups, to address these issues.

## Ethical Values in the Pharmaceutical Industry

The pharmaceutical industry plays a crucial role in the health and well-being of individuals. And with this important role comes the responsibility to adhere to a strict code of ethical values, such as ensuring the safety and efficacy of drugs, providing accurate and transparent information about drugs to consumers and healthcare professionals, and conducting clinical trials responsibly and ethically. Additionally, the industry is responsible for ensuring that drugs are available and affordable to those who need them and for refraining from engaging in illegal or unethical marketing practices, such as promoting their drugs for unapproved or off-label uses, luring the prospective prescribers with financial and other incentives to prescribe their products, and amplifying their product benefits and downplaying their side effects.

The pharmaceutical industries and the governments in most countries created ethical codes of practice indicating the ethical values, standards and practices that the drug companies should follow to curb corrupt and unethical practices. The following section presents a brief overview of the ethical codes applicable to pharmaceutical marketing worldwide.

## Ethical Codes in Pharma Marketing

The global pharmaceutical industry has made significant efforts towards ensuring compliant and ethical communication and interaction with physicians and patients over the past two decades.

The pharmaceutical industry must provide physicians and patients with the information to keep up with the ever-increasing treatment options and new products. In addition, the information companies provide should be scientifically accurate and fair, and their interactions with healthcare professionals are appropriate. The global pharmaceutical industry has made significant progress in providing objective information to support good patient care and meaningfully control companies' interactions with healthcare professionals with appropriate information about their products and services.

Information provided by pharmaceutical companies is of two types—promotional and non-promotional or scientific. However, the distinction between promotional and non-promotional may not always be clear. Promotional information encompasses advertising and sales material related to particular products that the companies distribute to patients through advertising campaigns or to healthcare professionals by pharmaceutical companies through their representatives. Non-promotional information focuses on the current understanding of certain diseases and treatment options not related to specific products. Scientific information mainly includes research contributions and developments concerning newer therapeutic options published in peer-reviewed journals.

In addition to the ethical standards and codes by pharmaceutical manufacturers' associations worldwide, governments too, in many countries employ a range of quality control mechanisms to evaluate the pharmaceutical industry's promotional communications and interactions with healthcare professionals. These mechanisms comprise industry codes of practice, internal company procedures, laws, regulations, and different governing mechanisms reflecting the national circumstances of the respective countries and regions. The following table summarizes global codes and regulatory mechanisms applied to pharmaceutical companies.

**Table 2.1 WHO Ethical Criteria for Medicinal Drug Promotion (1988)—Exerts Influence Over Company, HCPs and Government Codes, Regulations and Laws**

| International | Regional | National |
|---|---|---|
| IFPMA Code of Practice (First Established in 1981 Latest Revision in 2019) | EFPIA Code of Practice | National Code of Practice of IFPMA Affiliated Associations |
| | Federacion Centroamericana de Laboratorios Farmacéuticos FEDEFARMA | National Code of Practice of IFPMA Affiliated Associations |
| Regulatory | | |
| US FCPA<br>UK Bribery Act | European Directives | National Laws & Regulations |

(Source: Adapted from article—Jeffrey Francer, Jose Zamarriego Izquierdo, Tamara Music, Kirti Narsal, Ethical Pharmaceutical Promotion and Communications Worldwide: Codes and Regulations, Philosophy Ethics and Humanities in Medicines, 9 (1):7, ResearchGate, March 2014)

## Consensus Frame Work

A Consensus framework in pharma marketing ethical codes refers to a set of guidelines or principles that have been agreed upon by a group of stakeholders in the pharmaceutical industry, such as industry representatives, healthcare professionals, and patient groups. These guidelines are meant to provide a common understanding of what constitutes ethical behavior in the marketing and promotion of pharmaceutical products is and should be. The goal of a consensus framework is to establish a shared understanding of what is considered appropriate conduct and to guide companies, healthcare professionals, and other stakeholders in navigating complex ethical issues in the industry.

Thus, a consensus framework typically addresses topics such as transparency, interactions with healthcare professionals, the responsibility of sharing information, and the protection of patient privacy.

IFPMA Code is one framework that is widely accepted across the globe. It is a set of ethical guidelines for the pharmaceutical industry that promotes responsible and ethical behavior.

**Industry Codes of Practice**

Although various pharmaceutical industry codes of practice have existed for several decades, there has been a shift in the industry's attitudes toward communications activity prompting regular revisions to international codes with dependent national codes being updated since the beginning of 2002.

For example, in 2002, the Pharmaceutical Research and Manufacturers' Association of America (PhRMA) substantially updated its national code in the US. IFPMA (International Federation of Pharmaceutical Manufacturers & Associations) revised its code in 2006 and updated it again in 2012 to synchronize national efforts. IFPMA members, consisting of companies and national trade associations, must adopt its code. Companies that are not direct IFPMA members may also be bound to the same threshold requirements because of their relationship with IFPMA national associations. The IFPMA Code covers interactions with healthcare professionals and the promotion of pharmaceutical products.

**IFPMA Code: The Guiding Principles**

Over the years, the IFPMA Code of Pharmaceutical Marketing Practices has been revising and updating its code, simplifying the language of previous code, and expanding the rules relating to company interactions with healthcare professionals.

First adopted in the 1950s, the periodic updates in 2006 and 2012 have addressed the changing landscape globally to prevent potential incidents in business practices, such as interactions between pharmaceutical companies and healthcare stakeholders. Here is an illustrative list of the topics covered in the updated sections of the code:

**Table 2.2 List of Topics Covered in the Updated Sections of the IFPMA Code**

- Fundamental requirements for ethical and professional behavior, putting patients first, compliance with regulations, etc
- Standards for interactions between pharma companies and healthcare professionals
- Sponsorship or support for healthcare professionals' attendance at meetings and continuing medical education
- Acceptability of venues and locations for meetings
- Fees for service for engagement of healthcare professionals
- Providing promotional aids, samples, etc.
- Hospitality limitations
- Standards for promotional information— accuracy, balance, substantiation, etc.
- Essential prescribing information for advertisements (e.g., prescribing information)
- Prohibition of promotion of unlicensed products and uses
- Electronic communications
- Interactions with patient organizations
- Clinical research and transparency
- Company procedures and responsibilities, including approval and certification arrangements, staff training
- Complaints handling and enforcement arrangement

In addition, all European and some other nations covered more areas in their national codes, such as:

- Prohibition of Direct-to-consumer advertising for prescription-only medicines
- Specific requirements for representatives
- Requirements for public listings of support and engagement of healthcare professionals and patient groups
- Donations and Grants
- Non-interventional studies
- Aspects of market research activities
- Providing educational support services, such as therapy reviews and nurse reviews
- Standards for non-promotional medical information to healthcare professionals and patients
- Non-promotional information for patients and the public, such as disease awareness activities
- Interactions with media, press releases, etc.
- Specific requirements for websites, social media, etc

(Source: Adapted from an article— Ethical pharmaceutical promotion and communications worldwide: codes and regulations by Jeffrey Francer1, Jose Zamarriego Izquierdo 2, Tamara Music 3, Kirti Narsai 4, Chrisoula Nikidis 5, Heather Simmonds 6, and Paul Woods7, published in *Philosophy, Ethics, and Humanities in Medicine*, 2014, 9:7 on March 29, 2014.)

Generally speaking, the requirements of codes and legislation usually overlap extensively. For example, a promotional claim or an illegal activity will also breach the local code of practice. However, in many countries, the code requirements are broader than those in legislation and provide more detail on exactly what is applicable and what is not.

**Recent Updates to IFPMA Code**

First began in 1950, the IFPMA Code went through many revisions and expansions over the years, reflecting and extending its scope beyond marketing activities. For example, the 2012 revisions addressed issues such as fees for services, clinical research transparency, and interactions with patient organizations. In addition, the Code became more comprehensive, clarifying certain articles and extending its scope to issues such as the commissioning of advisory boards and support for continuing medical education (CME).

Particularly significant in the 2012 revision of the IPFMA Code is arriving at the *Guiding Principles*, which identify the underlying principles on which the rules had been based. These Guiding Principles are useful because no Code rules can cover all situations. Stating underlying principles should help interpret individual cases, and when IFPMA minimum standards are incorporated in more detailed local codes.

The Guiding Principles of the 2012 IFPMA Code of Practice are:

1. The healthcare and well-being of patients are the priority for pharmaceutical companies.
2. Pharmaceutical companies will conform to high quality, safety, and efficacy standards as determined by regulatory authorities.
3. Pharmaceutical companies' interactions with stakeholders must always be ethical, appropriate, and professional. Nothing should be offered or provided by a company in a manner or conditions that would have an inappropriate influence.
4. Pharmaceutical companies are responsible for providing accurate, balanced, and scientifically valid data on products.

5. Promotion must be ethical, accurate, balanced and must not be misleading. Information in promotional materials must support proper assessment of the risks and benefits of the product and its appropriate use.
6. Pharmaceutical companies will respect the privacy and personal information of patients.
7. All clinical trials and scientific research sponsored or supported by companies will be conducted with the intent to develop knowledge that will benefit patients and advance science and medicine. Pharmaceutical companies are committed to the transparency of industry-sponsored clinical trials in patients.
8. Pharmaceutical companies should adhere to applicable industry codes in the spirit and the letter. To achieve this, pharmaceutical companies will ensure that all relevant personnel are appropriately trained.

The four major areas that the 2012 revision of the IFPMA Code guidelines covered are:

- Companies' engagements with HCPs providing consulting services, such as scientific consulting, market research, and advisory board participation. The revised guidelines defined contractual relationships clearly and suggested a written contract or agreement defining a clear business need for the services provided, including a remuneration that reflects fair market value.
- Clinical research and transparency is another area that has been added to the Code, reflecting the long-standing commitment to disclose clinical trial information in line with the joint disclosures issued by international and national industry associations.
- Furthermore, a section on company support for continuing medical education (CME) has been added to the Code, reaffirming that the primary purpose of a CME is to enhance medical knowledge with contributions to content to be fair, balanced, and objective. The Code also added that similar provisions apply to promotional events, too.

- The revised Code also formulated rules governing the pharmaceutical companies' interactions with patient organizations to safeguard the patient organizations' independence and ensure that companies' support is appropriate. For example, written documentation must be in place regarding companies' involvement, and restrictions similar to HCP meetings apply to pharma companies' support to patient organizations.

The 2012 IFPMA Code edition clarified certain requirements, such as acceptable business practices and new standard operating procedures. For instance, pharmaceutical companies cannot provide entertainment and social activities to HCPs, whereas previous versions allowed modest entertainment.

Here are the main areas of coverage of international and national codes. However, in some countries, it is legislation rather than codes that cover certain requirements.

- Fundamental requirements for ethical and professional behavior, putting patients first, compliance with regulations, etc
- Standards for interactions between pharma companies and healthcare professionals
- Sponsorship or support for healthcare professionals' attendance at meetings and continuing medical education
- Acceptability of venues and locations for meetings
- Fees for service for engagement of healthcare professionals
- Providing promotional aids, samples, etc.
- Hospitality limitations
- Standards for promotional information— accuracy, balance, substantiation, etc.
- Essential prescribing information for advertisements, packaging inserts, etc.
- Prohibition of promotion of unlicensed products and uses
- Electronic communications
- Interactions with patient organizations

- Clinical research and transparency
- Company procedures and responsibilities, including approval and certification arrangements, staff training
- Complaints handling and enforcement arrangements

Further, European and certain other national codes cover certain additional areas, such as:

- Prohibition of Direct-to-consumer advertising for prescription-only medicines
- Specific requirements for representatives
- Requirements for public listings of support and engagement of healthcare professionals and patient groups
- Donations and grants
- Non-interventional studies
- Aspects of market research activities
- Providing educational and support services, such as therapy review and nurse services
- Standards for non-promotional medical information to healthcare professionals and patients
- Non-promotional information for patients and the public, and disease awareness activities
- Interactions with media, press releases, etc.
- Specific requirements for websites, social media, etc.

**IFPMA 2019: What is New?**

The International Federation of Pharmaceutical Manufacturers and Associations (IFPMA) revised its guidelines in 2019. Here's what is new in its latest edition:

- A strengthened section on transparency, including requirements for disclosing payments to healthcare professionals and organizations.
- More specific guidance on interactions between industry and healthcare professionals, including a prohibition on entertainment and gifts.

- New provisions on the responsible sharing of clinical trial data.
- Companies must have policies in place to address potential conflicts of interest.

Overall, the updates in the 2019 edition of the IFPMA Code are intended to increase transparency and accountability in the pharmaceutical industry and promote ethical interactions between the industry and healthcare professionals.

**Regulatory Mechanisms**

While most codes are voluntary to enable high standards without compromising effective communication from pharmaceutical companies that benefit healthcare providers and their patients, laws, and regulations to all parties are necessary to provide additional safeguards to discourage wrongdoing. For example, in Europe and the US, if an inappropriate payment or gift is given or offered by a company or requested or accepted by a healthcare professional, both parties could be penalized. Consider, for example, the following laws governing ethical and professional behavior and standards for interactions between pharmaceutical companies and healthcare professionals in the US:

- US Foreign Corrupt Practices Act (FCPA)
- Sunshine Act (US)
- UK Bribery Act

**Pharma Industry Interaction with HCPs**

Pharmaceutical advertising codes and legislation cover the rules governing the pharmaceutical industry's interactions and communication, as they are very important, even crucial, in the appropriate and effective use of prescription medicines. Furthermore, national bribery and corruption legislation, such as US Foreign Corrupt Practices Act (FCPA), and UK Bribery Act, could potentially apply to pharmaceutical companies' activities in any country. These codes help ensure compatibility with relevant sections of anti-bribery legislation prohibiting inappropriate personal benefits offered to healthcare professionals and often go beyond the requirements of anti-bribery legislation.

Most national Codes cover the issues concerning the companies' support of healthcare professional attendance at medical conferences. Codes in many countries consider sponsoring the attendance of HCPs by covering associated costs such as reasonable travel, accommodation, and meals at scientific meetings acceptable, albeit with certain caveats, such as:

- The meeting must be scientific and professional, and any refreshments provided must be incidental to that purpose.
- The venue must be conducive to scientific or educational purposes, and the international nature of the meeting or other logistical or security reasons must justify international travel.

However, company sponsorship of HCPs to attend meetings remains a topic of debate. Some countries, such as the United States and Norway, do not permit direct sponsorship of attendance at scientific meetings (except for medical students in the US). At the same time, countries like France require a review of the arrangements by an independent body. In addition, some countries have put other measures in place, such as co-payment of expenses. All these measures highlight sensitivity over policies relating to the sponsorship of healthcare professionals that go beyond external rules. However, ceasing sponsorship could deny HCPs access to hearing and interact with world leaders in their chosen field unless alternative funding arrangements are developed, or digitally based specialist educational services are expanded and are feasible in their country. This is particularly important for healthcare professionals from developing countries, where alternative funding sources may not be available.

While international rules still permit the use of inexpensive promotional aids, such as pens, pads, tongue depressors, antiseptic wipes, etc., provided they are relevant to the HCP's practice. However, there has been a trend to ban promotional aids altogether in some countries, such as the US and the UK, for the past few years. One global pharmaceutical company has even stopped its distribution of promotional aids worldwide. The rationale for a ban is not that such promotional aids represent a gift that will affect a healthcare professional's prescribing or purchasing decisions but rather that such items are not conducive to a new relationship built on mutual professional respect. Further, industry leaders seek to base relationships with HCPs on sharing educational information rather than providing items that could be perceived as gifts of gratification.

In most countries, it is permissible to provide samples of medicines to HCPs as they may improve patient care. However, the local industry codes restrict their number, frequency, and the period after launch during which they can be provided.

**Codes in Emerging Markets: An Overview**

It is important to have a quick overview of ethical codes governing the interactions of pharma and HCPs in emerging markets such as China, India, Latin America, and South Africa, as they are increasingly becoming the focus of international pharmaceutical companies.

**China**: In China, research-based international pharmaceutical companies represent a relatively small proportion of the total market. However, their presence is growing, and several companies have committed to major research and manufacturing investments there. RDPAC (R&D-based Pharmaceutical Association Committee), the trade organization representing international pharmaceutical companies in China, has a code of practice closely based on the IFPMA code. However, legal controls dominate, and the status of voluntary codes remains uncertain in China. For example, advertisements must be submitted to the Chinese regulatory authorities for approval before being issued.

**India**: Most pharmaceutical companies in India are national and do not operate in other countries. Several thousand companies are operating in only the domestic market, represented by national trade associations, such as IDMA (Indian Drug Manufacturers Association). While these associations have codes of practice, they are not bound by the standards and procedures set out by the international IPFMA code. At the same time, international companies, including some India-based companies that operate internationally, are members of OPPI (Organization of Pharmaceutical Producers of India) and are governed by its advertising code, which is closely linked to the IPFMA Code. Furthermore, in June 2011, the DOP (Depart of Pharmaceuticals) in India came up with the Uniform Code of Pharmaceutical Marketing Practices (UCPMP) to arrest the growing unethical marketing practices of bribing doctors by the pharma companies in India. The UCPMP Code is closely linked to the IFPMA Code. However, modern legislation governing pharmaceutical advertising is lacking, although the Drugs and Magic Remedies (Objectionable Advertisements) Act 1955 is still in force.

**Mexico:** In Mexico, a collaboration between the local and international pharmaceutical industry, medical associations, medical schools, government bodies, and others led to the agreement in 2008 of joint mandatory transparency guidelines. The Council of Ethics and Transparency (CETIFARMA) and the more restrictive standards for industry business conduct were brought out by the 2006 revision of the IFPMA Code. The Council is an autonomous and independent body that operates mandatory codes covering ethics and transparency, good promotional practices, and interaction with patient organizations. The Council monitors the codes and can apply sanctions where necessary. There is also a voluntary award system based on an independent evaluation of company compliance. Companies are certified for two years, after which they have to be evaluated again.

**South Africa:** In South Africa, the local industry associations, including manufacturers of medical devices, diagnostics, and generics, as well as prescription and over-the-counter medicines, have produced a joint code in line with legal provisions in the Medicines Act. Its implementation began in the autumn of 2011. Furthermore, an independent enforcement authority, The Marketing Code Authority, has been established under the Code to ensure its implementation with detailed enforcement procedures and the application of extensive and stringent sanctions in cases of code breaches. It also includes a certification process for industry professionals, which makes South Africa a good example of the willingness of all stakeholders to work together.

In addition, several emerging nations have made significant advances in regulating the advertising of prescription medicines in collaboration with government and healthcare professional organizations, which are worthy of emulation across developing and developed nations.

## Code of Practice Sanctions

Codes of practice operate on a fundamentally different basis from the legislation. They do not rely merely on the threat of punitive fines for their effectiveness, as they are voluntary and mutually agreed upon. Instead, codes represent a collective commitment of member companies to behave responsibly. Moreover, any deviations from the Code requirements are dealt with in ways consistent with local laws, including anti-trust and anti-competition provisions. The following table presents a summary of Code Practice Sanctions and Provisions.

**Table 2.3 Code of Practice Sanctions and Provisions: A Summary**

| Sanction or Requirement | Comment |
|---|---|
| Requirement to cease bribe-compliant activity | A universal requirement, often associated with a written compliant or similar activities, claims, etc. The company may be required to recover and destroy offending material. Repetition may result in severe penalties. |
| Publication of the outcome or public reprimand | Undertaken if local legal considerations allow. May consist of detailed reports or more concise summaries. Offending company is usually identified. In some countries, serious offenses may be publicized in the medical press.<br>The amount is usually graded according to the number and seriousness of the offenses, generally from thousands to hundreds of thousands of dollars. |
| Additional pre-screening requirements | In countries where pre-screening is optional |
| Requirements for a formal audit of company procedures | This is particularly useful if a company's procedures or training may be the cause of a serious repeated shortcoming |

Table 2.3 *Contd...*

| Sanction or Requirement | Comment |
|---|---|
| Suspension or expulsion from membership of the local trade association | Expulsion may mean that the code regulatory system will not apply to the company and external legal and regulatory controls will therefore take effect routinely. Suspension may mean that the company is still required to comply with the national association code. |
| Issue a corrective communication | This provision is particularly useful if recipients of the material may have been misled. It will be at the expense of the company. |

(Source: Adapted from an article— Ethical pharmaceutical promotion and communications worldwide: codes and regulations by Jeffrey Francer, Jose Zamarriego Izquierdo, Tamara Music, Kirti Narsai, Chrisoula Nikidis, Heather Simmonds, and Paul Woods, published in *Philosophy, Ethics, and Humanities in Medicine*, 2014, 9:7 on March 29, 2014.)

## Ethical Codes in LDCs

What is the current scenario of pharmaceutical marketing ethical codes in the Least Developed Countries (LDCs)? Generally speaking, most LDCs often lack regulatory oversight and enforcement of ethical codes in pharmaceutical marketing, leading to the following:

1. *Promotion of counterfeit drugs*: The distribution of fake or low-quality drugs that do not meet safety standards can harm patients and undermine public trust in the healthcare system.
2. *False Advertising*: Misleading or exaggerated claims about the benefits and risks of drugs can lead to over-diagnosis, over-treatment, and increased healthcare costs.
3. *Bribing healthcare professionals*: Pharmaceutical companies may offer bribes or incentives to healthcare professionals to prescribe their drugs, even if they are not the best choice for the patient.
4. *Kickbacks*: Pharmaceutical companies may pay kickbacks to suppliers, distributors, or other intermediaries to secure favorable treatment for their products.
5. *The exploitation of vulnerable populations*: In some cases, pharmaceutical companies may target vulnerable populations, such as low-income or rural communities, with substandard or unnecessary drugs.

6. *Failure to disclose conflicts of interest:* Pharmaceutical companies may not disclose their financial relationships with healthcare professionals, which can compromise the impartiality of their recommendations.

Since these practices can have serious consequences for public health and undermine the integrity of the healthcare system, it is important for governments and healthcare organizations, including pharmaceutical companies, to take steps to prevent and address corrupt and unethical practices in pharmaceutical marketing.

### Are Marketing Ethical Codes Mere Paper Tigers?

Pharmaceutical marketing ethical codes, however well-meaning they are in promoting responsible and ethical practices in the industry, cannot be effective if they are not backed by adequate regulatory oversight and strong enforcement mechanisms.

In some countries, there is a well-developed regulatory framework and strong enforcement mechanisms, which can help to ensure that ethical codes are followed and that the violators are held accountable. Whereas in other countries, where the regulatory environment may be inadequate, it is easier for unethical practices to occur. As a result, pharma marketing ethical codes remain in those countries as mere paper tigers.

### What Can Make Ethical Codes Work?

The incentives that drive industry behavior may limit the effectiveness of ethical codes. For example, companies may prioritize maximizing profits over following ethical standards when the financial or legal consequences are limited. What is needed, therefore, is a combined and concerted effort by governments, healthcare organizations in general and pharmaceutical companies in particular, and other stakeholders to strengthen regulatory frameworks, improve enforcement mechanisms, and promote a culture of ethical responsibility in the industry.

CHAPTER

3

# Unethical Marketing Practices in Pharma

No one should approach the temple of science with the soul of a money changer.

*—Thomas Browne, English Physician and Author, Best Known for his book, Reflections*

## Unethical Marketing Practices in Pharma

Unethical marketing practices in the pharmaceutical industry are a major concern worldwide. These practices can range from many areas, such as:

- Providing false or misleading information about drugs.
- Bribing HCPs to prescribe certain medications. In addition to being unethical, such a practice can lead to overusing certain drugs and create a conflict of interest for HCPs.
- Promotion of drugs for unapproved uses, known as *off-label* promotion. This occurs when a pharma company markets a medication that has not been approved for use in that condition by regulatory agencies such as the FDA. This can be dangerous for patients as it may lead to the use of a drug that is not effective or safe for its intended use.
- Ghostwriting, where a pharmaceutical company hires a third-party company to write research papers or articles that promote their products, but the authorship is attributed to academic or industry experts who have not necessarily written the content. This practice misleads the readers and the public by giving them the impression that an independent expert has written the content.

Additionally, direct-to-consumer advertising (DTCA) is controversial and considered an unethical marketing practice. However, it is approved by the regulatory authorities in only a few countries, such as the US and New Zealand. Furthermore, critics of DTCA argue that it can lead to patients demanding unnecessary drugs from their physicians and may create unrealistic expectations about the benefits and risks of a particular medication.

There have been numerous examples of pharmaceutical companies engaging in unethical practices worldwide. Here are some notable examples:

- In 2012, GlaxoSmithKline (GSK) was fined $3 billion for illegal marketing practices and failure to report safety data for some of its drugs. The company was found to have promoted the antidepressant Paxil for unapproved uses, including treating children and teenagers, downplaying the risks associated with the drug.

- In 2013, Johnson & Johnson was ordered to pay $2.2 billion in fines for promoting the anti-psychotic drug Risperdal for unapproved uses, including treating elderly patients with dementia. The company was also found to downplay the risks associated with the drug.
- In 2016, Novartis AG was ordered to pay $390 million in fines for bribing doctors in Greece to prescribe its drugs. The company was found to have provided kickbacks to doctors through cash, gifts, and rebates.
- In 2020, Purdue Pharma, the manufacturer of OxyContin (Oxycodone), pleaded guilty to three federal criminal charges for its role in the opioid epidemic. The company admitted to knowingly and intentionally misbranding the drug as less addictive and risky than it was.
- In 2020, Teva Pharmaceuticals, one of the largest generics manufacturers in the world, agreed to pay $85 million to settle charges that it bribed foreign officials to win business.

These are only a few examples demonstrating the range of unethical practices in the pharmaceutical industry, from illegal marketing to bribing healthcare providers and government officials. In reality, unethical practices are much more rampant than these. Therefore, governments and regulatory authorities should pay serious attention and make legislative changes to curb kickbacks and corruption. In addition, regulatory agencies must remain vigilant to prevent illegal marketing practices.

The following case studies indicate the extent of unethical marketing practices that have been taking place in pharmaceutical marketing worldwide and how the respective governments have been trying to curb those.

CASE

5

# What Wrong Did GSK Do to Pay A Whopping $3 billion in Settlement Fees?!

On July 2, 201, GlaxoSmithKline agreed to a criminal settlement with the Department of Justice to settle accusations that it didn't tell regulators about safety problems with an infamous drug, marketed other medicines for diseases they were not approved and might not have been safe and followed marketing practices that were not legal and ethical to increase their sales.

**What Wrong Things Did GlaxoSmithKline Do?**

The company pleaded guilty to misdemeanors, accounting for about $1 billion of the settlement. Consider, for example some of these:

1. GSK failed to disclose safety data from certain studies of Avandia (Rosiglitazone Maleate, an oral anti-diabetic) conducted between 2001 and 2007 to the US FDA. This is, ethically, perhaps the most serious of the charges. Glaxo's holding of the Avandia data was a bad disclosure, bordering on deceit. During that period, Avandia became the best-selling diabetes drug in the world. Now, after the investigation, Avandia not only bears warnings that it might cause heart attacks, its use has been so restricted that the drug has nearly vanished from the company's ledgers. Moreover, Glaxo's concealing of the heart risk data meant that patients were exposed to added risks, for which GSK paid a fine of $243 million.

2. Glaxo also pleaded to misdemeanor criminal charges that it sold two antidepressants— Paxil (Paroxetine) and Wellbutrin (Bupropion) for two unapproved indications. Glaxo promoted Paroxetine to treat depression in children and Wellbutrin for treating sexual dysfunction, obesity, and ADHD (Attention Deficit Hyperactivity Disorder), all of which were not approved. With Paxil and Wellbutrin, Glaxo was accused of paying experts to speak at meetings to other doctors about research on these uses, even though the FDA had not approved those uses. The United States contention was that GSK paid millions of dollars to doctors to speak at and attend meetings, sometimes at lavish resorts, at which the off-label uses of Wellbutrin were routinely promoted and also used sales representatives, sham advisory boards, and supposedly independent Continuing Medical Education (CME) programs to promote Wellbutrin for these unapproved uses. These meetings amplified scientific arguments for the drug's use and drowned out critical voices. These drugs are still marketed and used for their approved indications, but GSK paid a $757 million fine for violations, such as promoting them in unapproved indications.
3. In addition, GlaxoSmithKline had to pay $2 billion to settle civil charges, including more allegations of off-label marketing, of Avandia and Advair (Fluticasone and Salmeterol), the anti-asthmatic drug, Lamictal (Lamotrigine), the anti-epileptic drug and Zofran (Ondansetron) the antiemetic (approved only for use with cancer chemotherapy, but according to the FDA the company marketed it to pregnant women with morning sickness).

(Source: Adapted from an article— *The Terrible Things GlaxoSmithKline Did Wrong— And the Thing It's Doing Right*, published in Forbes.com on July 2, 2012)

CASE

6

# OxyContin Marketing: Commercial Triumph, Public Tragedy!

Purdue Pharma, a privately held pharmaceutical company from Connecticut, US, introduced OxyContin in 1995.

OxyContin is a branded, extended-release oxycodone tablet approved for managing pain severe enough to require daily, around-the-clock, long-term opioid treatment for which alternative treatment options were inadequate. Therefore, it is classified as a Schedule II narcotic under the Controlled Substances Act.

## Commercial Triumph

In 1996, its launch year, OxyContin recorded $45 million in sales for Purdue. By 2000, OxyContin became a blockbuster drug, generating $1.1 billion in sales. Further, in 2010, profits from OxyContin rose to $3.1 billion.

The Sackler family, including Mortimer, Raymond Sackler, and their brother Arthur purchased Purdue Pharma in 1952. While the Sackler brothers made their original fortune in medical advertising, OxyContin sales' launch and phenomenal growth have significantly increased the family's wealth. In 2015, Forbes estimated the Sackler family wealth at $14 billion, shared among twenty family members.

## Aggressive Marketing of OxyContin

What did Purdue do with OxyContin's promotion to make it such a profitable product?

Firstly, the company expanded its sales force after debuting OxyContin in 1996. Purdue had more than 300 sales representatives in 1996 when it launched OxyContin. The company had more than doubled its sales force to over 600 by 2000, and further to more than 1,000 sales representatives by 2010. Physician coverage too increased from 33,400 in 1996 to 94,000 in 2000.

Secondly, the company incentivized the sales representatives with large bonuses, cash prizes, and luxury vacations. For example, in 2001, annual bonuses for sales reps averaged more than $70,000, with some earning nearly $250,000 in bonuses. Furthermore, the management strongly persuaded the sales force to promote stronger doses to physicians in their conversations. This is evident from a sales manager's memo to her team titled, "It's Bonus Time in the Neighborhood!" The manager wrote in the memo that he, who sold 40 mg (the largest dose at the time), would win the battle. Internal memos referred to sales teams as *Crusaders* and *Knights* and executives as the Royal Court of OxyContin to increase motivation levels.

Thirdly, Purdue and its sales team convinced doctors of the safety of OxyContin. Most doctors have been reluctant to prescribe opioids because they are addictive and not meant to be used for chronic pain. How did Purdue change this reluctance to prescribing into prescribing OxyContin prolifically? Some of the key marketing strategies and tactics that Purdue Pharma implemented include speaker-training programs, profiling prescribers, targeting primary care physicians, targeting non-malignant pain market and downplaying the risk of addiction.

In addition, Purdue used a patient starter coupon program for OxyContin by providing patients with a free limited-time prescription for a 7-to-30-day supply. By 2001, when the program was ended, around 34,000 coupons had been redeemed nationally.

The company also followed an unprecedented promotional campaign for a controlled substance drug and a schedule II Opioid, distributing branded promotional items, such as OxyContin fishing hats, stuffed plush toys, and music compact discs (Get in the Swing with OxyContin).

**Speaker-Training Programs**

Purdue trained over 5,000 physicians, pharmacists, and nurses as effective pain-management speakers through all-expenses-paid symposia and recruited them for the company's national speaker bureau. For example, the company conducted more than forty national pain-management speaker-training conferences at resorts in Florida, Arizona, and California from 1996 to 2001. Symposia like these influence physicians' prescribing, even though the physicians who attend such symposia believe such enticements do not alter their prescribing patterns.

**Profiling Prescribers**

Another major contributing factor to Purdue's marketing success is using sophisticated marketing data to influence physicians' prescribing. For example, it is a common practice for pharmaceutical companies to compile prescriber profiles on individual physicians nationwide to influence doctors' prescribing habits. As a result, a pharmaceutical company can identify the highest and lowest prescribers of particular drugs in a single zip code, country, state, or entire country. In addition, Purdue targeted the highest number of opioid-prescribing physicians nationwide and targeted them to promote OxyContin aggressively. Thus, Purdue had a clear idea of which physicians had large numbers of chronic pain patients and who had the least numbers.

**Targeting Primary Care Physicians**

Purdue targeted primary care physicians and promoted a more liberal use of opioids, significantly expanding the use of particularly sustained-release opioids. As a result, by 2003, nearly half of all primary care physicians in the US were prescribing OxyContin. However, many experts were concerned about OxyContin usage by primary care physicians as they had the least time to evaluate and follow up with patients with complicated chronic pain, particularly in a managed care environment.

### Targeting Non-malignant Pain Market

Non-malignant pain market is much larger than the cancer-related pain market, and Purdue aggressively promoted OxyContin in the non-malignant pain market. As a result, the non-cancer-related pain market constituted 86 percent of the total opioid market in 1999. Furthermore, Purdue's aggressive promotion of OxyContin for treating non-cancer-related pain contributed to a ten-fold increase—from about 670,000 in 1997 to 6.2 million in 2002 for OxyContin prescriptions for this type of pain. In contrast, prescriptions for cancer-related pain increased only about fourfold.

### Downplaying the Risk of Addiction

Purdue consistently downplayed the risk of possible addiction in the use of opioids for the treatment of non-cancer-related pain. However, one of the most critical issues regarding the use of opioids in the treatment of chronic non-cancer-related pain is the potential for iatrogenic addiction. But, in much of its promotional campaign, including literature, audiotapes for physicians, brochures and videotapes for patients, and its "Partners Against Pain" website, Purdue claimed that the risk of addiction from OxyContin was extremely small. In addition, the company trained its sales representatives to communicate to physicians that the risk of addiction was less than one percent, substantiating this claim with two studies — a study by J. Porter and H. Jick titled *Addiction Rare in Patients Treated With Narcotics* published in New England Journal of Medicine in 1980, and the other titled *Management of Pain During Debridement: A survey of US Burn Units* by Perry S, and Heidrich G published in Pain journal in 1982.

However, several studies demonstrate that in treating chronic non-cancer-related pain with opioids, there is a high incidence of prescription drug abuse.

**High Abuse Potential**

Drug abusers can crush the controlled-release tablet and swallow, inhale, or inject the high-potency opioid for an intense morphine-like high. Some precedence existed for the diversion and abuse of controlled-release opioid preparations. For example, Purdue's MS Contin (Morphine Sulphate Sustained-Release) tablets had been abused in the late 1980s as it is now since the 1990s with OxyContin. Purdue's own testing in 1995 had demonstrated that 68 percent of Oxycodone could be extracted from an OxyContin tablet when crushed. As a result, the great commercial success of OxyContin was stained by increasing rates of abuse and addiction, as was the case with MS Contin earlier in the 1990s.

With increasing prescriptions of OxyContin, the non-medical use of OxyContin also has grown considerably. There seems to be a strong correlation with the commercial success of OxyContin, fueled by an unprecedentedly aggressive promotion and marketing campaign that was stained by escalating OxyContin abuse and diversion spreading throughout the United States. By 2004, OxyContin was associated with higher abuse rates, becoming the most prevalent abused prescription opioid in the US.

**Huge Promotion Spend**

Purdue spent an extraordinary amount of money promoting a sustained-release opioid, OxyContin. In its first six years on the market, the company spent approximately 6 to 12 times more on promoting it than it had spent on promoting MS Contin or Janssen Pharmaceutical Products had spent on Duragesic, one of OxyContin's competitors. As a result of such a massive promotion effort, OxyContin became the most frequently prescribed brand-name opioid in the United States for treating moderate to severe pain by 2001. However, it has not been shown to be superior to other available potent opioid preparations.

In addition, the United States General Accounting Office (GAO) stated in its report on *Prescription Drugs: OxyContin Abuse and Diversion and Efforts to Address the Problem* to Congressional Requesters, Purdue funded more than 20,000 pain-related educational programs through direct sponsorship or financial grants, providing a venue that enormously influenced physicians' prescribing throughout the country.

### Co-Promotion With Abbott

Between 1999 and 2002, Purdue focused on marketing OxyContin to specialists, such as pain management doctors, and entered into a co-promotion agreement for OxyContin with Abbott Laboratories to expand its reach and coverage to primary care physicians and other healthcare providers.

While the co-promotion agreement between the two companies is not illegal, it has been criticized for contributing to the opioid epidemic. The main allegation against the co-promotion arrangement was that it helped increase the number of prescriptions for OxyContin. In addition, the two companies' coordinated marketing efforts helped create a false perception of the drug's safety and efficacy. It is worth noting that these accusations and allegations are still under investigation, and the outcome of the cases will depend on the facts and evidence presented in each case.

In addition, Purdue used a patient starter coupon program for OxyContin by providing patients with a free limited-time prescription for a 7-to-30-day supply. By 2001, when the program was ended, around 34,000 coupons had been redeemed nationally.

The company also followed an unprecedented promotional campaign for a controlled substance drug and a schedule II Opioid, distributing branded promotional items, such as OxyContin fishing hats, stuffed plush toys, and music compact discs (Get in the Swing with OxyContin).

**OxyContin: Making of A Blockbuster**

Abbott is a much larger company than Purdue, with a sales force entrenched in hospitals and surgical centers, and has existing relationships with anesthesiologists, emergency room doctors, surgeons, and pain management teams. Abbott devoted 300 sales reps to OxyContin sales— about the same number of people Purdue initially dedicated to the drug as a part of their co-promotional agreement with Purdue.

Abbott marketed OxyContin from 1996 to 2002, right after the US FDA approved the drug, and helped Purdue achieve a blockbuster status for OxyContin from a mere $49 million in 1996 to over $1.6 billion in 2002. Over the life of the partnership, Purdue paid Abbott nearly a half-billion dollars, according to court orders.

How Did Abbott help Purdue achieve the sales success that it did?

- Heavily incentivized its sales staff to push OxyContin, offering $20,000 cash prizes and luxury vacations to top performers.
- Used the terminology from Middle Ages *Crusaders* to create hype and motivate the sales force. For example, they called the sales reps *Royal Crusaders* and *Knights* in internal communications and supervisors the *Royal Court of OxyContin*— and referred executives as the *Wizard of OxyContin*, *Supreme Sovereign of Pain Management*, and *Empress of Analgesia*. The head of pain care sales, Jerry Eichhorn, was the *King of Pain* and signed internal office documents simply as *King*.
- In a memo, Abbott urged its OxyContin sales force to continue their OxyContin banner onto the battlefield and keep highlighting OxyContin benefits to their doctors.
- The company also told its reps to highlight the drug's less abuse and less addiction potential, which could be taken twice a day because of its time-release formulation.

- Another method of promoting OxyContin was *Dine and Dash*, where the company sales reps would pay for takeout food at a restaurant favored by a particular doctor. During the five minutes it takes for the physician to pick up the food and the sales rep to pay for it, the rep would detail the benefits of OxyContin to the doctor.
- Other techniques and tactics involved inviting a surgeon to a bookstore, giving the doctor a coupon, and then detailing the doctor while waiting to pay.
- In addition to bringing lunches to doctor's offices, Abbott reps were encouraged to schedule afternoon cookie and candy snacks for those offices to build goodwill with appreciative staff.
- The creative tactics and communication that helped the company beat sales targets for OxyContin included earning tickets to One-Eyed Jack Lottery ticket rewards to all Crusaders in 2000. According to a company's sales memo, the tickets were worth up to $20,000. Furthermore, sales reps who achieved the greatest growth in market share were promised trips to *a location befitting questing and conquering Crusader*!

According to a company executive's statement in a 1997 document, the prescriptions written by Abbott MDs comprised 25 percent of OxyContin prescriptions. In addition, from late 1996 through 2002, Abbott received $374 million in commissions on OxyContn sales. Total sales of the drug during that time were nearly $5 billion.

Also, the more Abbott generated in sales, the higher the reward for the company. as per the agreement with Purdue, Abbott received 25 percent of all net sales, up to $10 million, for prescriptions written by doctors its sales reps called on, and 30 percent of sales above $10 million, according to court records.

Additionally, although Abbott stopped selling OxyContin in 2003, the company still received a residual payment of 6 percent of the net sales through 2006, per the West Virginia court records.

However, Purdue indemnified Abbott from the lawsuits, and the indemnification saved Abbott from legal costs and unflattering publicity.

**What went wrong with Purdue's Marketing Strategy?**

Purdue Pharma has followed a very aggressive strategy for promoting OxyContin, often overlooking many ethical concerns and practices, such as misleading communication, underplaying side effects and addiction potential, and paying kickbacks to physicians and healthcare providers for prescribing OxyContin. Here is an indicative list of accusations and allegations that the company has done:

- Overstated the drug's effectiveness for treating chronic pain, an indication for which it is not approved
- Downplayed the risk of addiction associated with OxyContin
- Claimed that the drug was less addictive and less subjective to abuse than other pain medications
- Targeted vulnerable populations, such as elderly patients and veterans, with its marketing
- Encouraged doctors to prescribe higher doses of OxyContin, even for patients with mild pain
- Misrepresented the drug's safety profile in its marketing materials
- Paid kickbacks to doctors and other healthcare providers who prescribed OxyContin

Indiscriminate and unethical marketing practices such as these led to increased addiction to opioid substances, including OxyContin, and reached an epidemic proportion.

It is also alleged that encouraged by the success of these marketing strategies, Purdue has also tried to replicate them in other international markets. As a result, Purdue started facing several lawsuits in many cities, states, and even countries seeking damages for the costs associated with the opioid epidemic.

### The settlement with the Department of Justice (DOJ)

In 2007, Purdue Pharma reached a settlement with the United States Department of Justice (DOJ) for offenses related to the marketing and sale of OxyContin. The company pleaded guilty to misbranding OxyContin as less addictive and less subjective to abuse than other pain medications and paid over $600 million in fines. Three company executives also pleaded guilty to criminal charges and were ordered to pay fines and serve probation. However, the settlement did not include any admission of liability by Purdue Pharma or the individual defendants.

### Current Status

What is the current status of claims on Purdue's OxyContin Cases? As of 2021, Purdue Pharma and its owners, the Sackler family, have faced numerous claims and lawsuits related to their marketing and sale of OxyContin. In 2019, Purdue filed for bankruptcy as part of a settlement agreement that included the company paying $10 billion to resolve claims related to its role in the opioid epidemic. However, the settlement is not finalized, as many legal charges remain pending. Therefore, it is uncertain how these pending cases will be resolved.

As of 2021, The Sackler Family is now facing lawsuits and investigations in many countries, especially in the UK and Canada. As a result, many Museums and institutions have refused to accept any more donations from them.

(Source: Adapted from articles— (1) Art Van Zee, MD, The Promotion and Marketing of OxyContin: Commercial Triumph, Public Health Tragedy, American Journal of Public Health, 2009 February; 99(2): 221-227, (2) Shraddha Chakradhar and Cassey Ross, The History of OxyContin, Told through Unsealed Purdue Documents, STAT, December 3, 2019, (3) United States General Accounting Office Report to

Congressional Requesters, Prescription Drugs: OxyContin Abuse and Diversion and Efforts to Address the Problem, GAO (4) David Armstrong, Secret Trove Reveals Bold 'Crusade' to Make OxyContin A Blockbuster, STAT, September 22, 2016)

CASE

7

# Big Pharma's Big Fines!

The use of unethical and illegal marketing practices has been widespread in the pharmaceutical industry for some time now. Unfortunately, the extent of violations, too, has been considerable, and even top pharmaceutical companies have been engaging in these to improve sales and gain market share. For example, in the five years between 2009 and 2014, eleven of the big pharma companies paid a whopping $13.7 billion in fines to resolve criminal and civil allegations by the Department of Justice in the United States for several violations, including off-label promotion of prescription drugs for unapproved conditions, misbranding and providing misleading information about safety and kickbacks to increase sales to hospitals.

Between 2009 and 2014, Pharmaceutical companies agreed to pay over $13 billion to resolve US Department of Justice Allegations of unethical and illegal marketing practices, including the promotion of medicines for uses that the Food And Drug Administration did not approve. Here are the details:

1. *Pfizer*: In September 2009, the US Department of Justice fined the company for misbranding Bextra (Valdecoxib) with the intent to defraud or mislead, promoting the drug to treat acute pain at dosages the FDA had previously deemed dangerously high. Bextra was pulled from the market in 2005 due to safety

concerns. Further, the government alleged that Pfizer also promoted three other drugs illegally: the antipsychotic Geodon (Ziprasidone), an antibiotic Zyvox (Linezolid), and the antiepileptic Lyrica (Pregabalin).

2. *Eli Lilly*: In January 2009, Eli Lilly was fined $1.42 billion to resolve a government investigation into the off-label promotion of the antipsychotic Zyprexa (Olanzapine). Zyprexa had been approved for treating certain psychotic disorders, but Lilly admitted to promoting the drug in elderly populations to treat dementia. The government also alleged that Lilly targeted primary care physicians to promote Zyprexa for unapproved uses and *trained its sales force to disregard the law*.
3. *AstraZeneca*: In April 2010, AstraZeneca was fined $520 million to resolve allegations that it illegally promoted the antipsychotic drug Seroquel (Quetiapine). Seroquel had been approved for treating schizophrenia and later for bipolar mania, but the government alleged that AstraZeneca promoted Seroquel for various unapproved uses, such as aggression, sleeplessness, anxiety, and depression. AstraZeneca denied the charges but agreed to pay the fine to end the investigation.
4. *Merck*: Merck agreed to pay a fine of $950 million related to the illegal promotion of the painkiller Vioxx (Rofecoxib) in 2011. Vioxx was withdrawn from the market after studies found the drug increased the risk of heart attacks. The company pled guilty to having promoted Vioxx as a treatment for rheumatoid arthritis before it had been approved for that use. The settlement also resolved allegations that Merck made false or misleading statements about the drug's heart safety to increase sales.
5. *Abbott*: In 2012, Abbott was fined $1.5 billion in connection to the illegal promotion of the antipsychotic drug Depakote (Divalproex). Abbott admitted to having trained a special sales force to target nursing homes, marketing the drug to control aggression and agitation in elderly dementia patients.

Depakote had never been approved for dementia, and Abbott lacked evidence that the drug was safe or effective in those conditions. The company also admitted to marketing Depakote to treat schizophrenia, even though no study had found it effective in schizophrenia.

6. *GlaxoSmithKline*: GlaxoSmithKline agreed to pay a fine o $3 billion fine to resolve old and criminal liabilities regarding its promotion of drugs and its failure to report safety data. This is the largest healthcare fraud settlement in the United States to date. The company pled guilty to misbranding the drug Paxil (Paroxetine) for treating depression in patients under 18, even though the drug had never been approved for that age group. GlaxoSmithKline also pled guilty to failing to disclose safety information about the diabetes drug Avandia (Rosiglitazone) to the FDA.
7. *Boehringer Ingelheim*: In October 2012, Boehringer Ingelheim Pharmaceuticals agreed to pay $95 million to resolve allegations that the company promoted several drugs for non-medically accepted uses. These drugs included the stroke prevention drug Aggrenox (Aspirin and Dipyridamole), the lung disease drugs Atrovent (Ipratropium Inhalation), Combvent (Ipratropium and Albuterol), and Micardis (Telmisartan), a drug to treat high blood pressure. In addition, the FDA alleged that Boehringer improperly marketed the drugs and *caused false claims to be submitted to government healthcare programs*.
8. *Sanofi-Aventis*: Sanofi-Aventis agreed to pay $109 million in December 2012 to resolve allegations that the company gave doctors free units of Hyalgan (Hyaluronate), an injection to relieve knee pain to encourage those doctors to buy their product. Sanofi lowered the effective price by promising these free samples to doctors, and at the same time, got inflated prices from government programs by submitting false price reports, alleged the United States government. According to the Department of Justice announcement, Medicare and other

government healthcare programs paid millions of dollars in kickback-tainted claims for Hyalgan.

9. *Amgen*: In December 2012, Amgen agreed to pay a $762 million fine to resolve criminal and civil charges that the company illegally introduced and promoted several drugs, including Aranesp (Darbepoetin Alfa), a drug to treat anemia. Amgen pleaded guilty to illegally selling Aranesp to be used at doses that the FDA had explicitly rejected and for an off-label treatment that had never been FDA-approved.
10. *Johnson & Johnson*: In November 2013, Johnson & Johnson agreed to pay a fine of $2.2 billion to resolve criminal and civil allegations relating to the prescription drugs Risperdal (Risperidone), Invega (Paliperidone) and Natrecor (Nasiritide). The government alleged that J&J promoted these drugs for uses not approved as safe and effective by the FDA, targeted elderly dementia patients in nursing homes, and paid kickbacks to physicians and the nation's largest long-term care pharmacy provider, Omnicore Inc. As part of the agreement, Johnson & Johnson admitted that it promoted Risperdal for treating psychotic symptoms in non-schizophrenic patients, although the drug was approved only to treat schizophrenia.
11. *Endo*: Endo Health Solutions Inc. and its subsidiary Endo Pharmaceuticals inc. agreed to pay $192.7 billion in February 2014 to resolve criminal and civil liability arising from Endo's marketing of the prescription drug Lidoderm (Lidocaine Transdermal Patch). As part of the agreement, Endo admitted that it intended that Lidoderm be used for unapproved indications and that it promoted Lidoderm to healthcare providers this way.

(Source: Lena Groeger, Pro Publica, Big Pharma's Big Fines, Propublica.org, February 24, 2014, and The Department of Justice, United States)

CASE

8

# Corrupt Marketing Practices in Vietnam's Pharmaceutical Industry!

Corruption, the abuse of entrusted power for private gain, is a pervasive problem affecting many parts of the world, including the pharmaceutical sector. The pharmaceutical sector causes a major obstacle to access to quality medicines by inflating pharmaceutical prices and wasting valuable resources. The pharmaceutical value chain has many potential targets for corruption, such as manufacturing, registration, medicine selection, procurement, distribution, prescribing, and dispensing.

In late 2004, WHO (World Health Organization) initiated the Good Governance for Medicines Programme to reduce corruption in the pharmaceutical systems by applying transparent, accountable administrative procedures and promoting ethical practices.

The opening of Vietnam in 1989 to foreign trade and the liberalization of rules governing pharmaceutical manufacturing, sale, and distribution transformed Vietnam's pharmaceutical supply chain from a centrally controlled system of a few State-owned pharmaceutical enterprises to a market-oriented system with the participation of a vast array of companies. They included 180 domestic pharmaceutical companies— with 22 foreign direct investment (FDI) producers, 90 importers, 800 domestic wholesalers and distributors, and 438 representative offices of foreign pharmaceutical companies by 2008.

Several researchers, including Tuan A. Nguyen, Rosemary Knight, Andrea Mant, Husna Razee, Geoffrey Brooks, Thu H. Dang, and Elizabeth E. Roughead, wrote an article on *Corruption Practices in Drug Prescribing in Vietnam*— an analysis based on qualitative interviews in BMC Health Services Research in 2018. Here is a gist of that article, showing how that 40 to 60 percent of the price of off-patent medicines in Vietnam was typically spent to induce prescribers to use the medicines and to persuade procurement officers within hospitals to buy them.

Vietnam's pharmaceutical market heavily depended on imported medicines accounting for more than 50 percent of the market share in 2011. Although domestic production of pharmaceuticals reached a 50 percent level at about the same time, over 90 percent of the raw materials used in domestic production were imported.

Vietnam's transition from a socialist economy to a market-based economy has presented many challenges, and surging prices of medicines were one of them.

The researchers undertook a qualitative study interviewing representatives from two groups in Vietnam's pharmaceutical industry— pharmaceutical companies and private pharmacies who set their medicine prices; and government officials responsible for controlling medicine prices in Vietnam.

The researchers presented a theoretically informed analysis of the interview data in a framework comprising three main groups— opportunities, pressures, and rationalization. Opportunities include monopoly, discretion, accountability, citizen voice, transparency, and enforcement. Pressures include wages, incentives, and pressure from clients. Finally, rationalization includes social norms, moral and ethical beliefs, attitudes, and personality. Here is a summary of the findings of the study:

1. *Opportunity for Corruption*: Opportunities for inducements in the Vietnam healthcare sector were mainly related to poor governance, which included:

- The degree of personal discretion afforded participants,
- Lack of transparency.
- Lack of accountability in administrative procedures, and
- A low probability of detection and follow-up of enforcement

2. *Discretion*: Discretion refers to the autonomous power of officials to make decisions that lead to improper decisions, such as medicine tendering practices in public hospitals and prescribing practices.

   This has resulted in kickbacks ranging from 3 to 30 percent of the sale to procuring authorities in the hospitals, out of which doctors got a small share as they did not have much say in the matter.

   However, the discretion of prescribers was manifested by their freedom to prescribe medicines in hospital outpatient consulting rooms. By contrast, in hospital inpatient wards, prescribed medicines were limited to the hospital formulary list and the availability of medicines in hospital pharmaceutical departments. Therefore, prescribing medicines in return for commissions was reported to be more prevalent in consulting rooms than inpatient wards. Therefore, the commission for prescribers in the consulting room was also often higher, up to 65 percent of the sale. In comparison, the commission for prescribers in inpatient wards was only around 15 to 35 percent.

   Furthermore, information on *who prescribed what and how many* was often recorded and provided to pharmaceutical companies by a nurse in each inpatient ward in exchange for a monthly salary of $ 10 to 15. Study participants from pharmaceutical companies also reported that they could double-check the prescribing information self-provided by their *contracted* prescribers (also called collaborators) with the hospital pharmacy staff who ran a service of counting prescriptions for a fee of $10 per product per month. Study participants from the doctors practicing in teaching hospitals pointed out that their prescribing freedom was limited. They often had to think twice before prescribing because their students always observed their behavior.

3. *Transparency*: Lack of transparency is one of the weakest points in governance, and Vietnam is no exception. For example, one of the participants, a manager from a State-owned pharmaceutical company in Vietnam, commented that *all financial activities are done with cash in hand, so we cannot control corrupt practices*.
4. Accountability: According to Derick W. Brinkerhoff of Research Triangle Institute, Washington DC, USA, accountability refers to the obligation of individuals or agencies to inform other actors about their decisions and actions, to justify or explain them, and to suffer sanctions and punishment for non-performance, misconduct or corrupt behavior. The first two components of accountability (providing information about what was done and why) form answerability which is the essence of accountability. The expected magnitude of any applicable punishment and sanctions for misconduct *gives teeth to accountability*. Poor answerability in Vietnam's health sector created more autonomous prescriber power, encouraging prescribers to order unnecessary medicines for their private gain. For example, a study participant stated that *a doctor prescribes four tonics in one prescription. That is because patients don't know what they are for. If patients knew about this, how can the doctor prescribe this? We don't have to explain what medicines are used for what purposes, so patients still have to buy all medicines prescribed*.
5. *Enforcement*: Brinkerhoff also stated that *sanctions without enforcement significantly reduce accountability*. Enforcement relies on the probability of detection of corrupt acts and the probability that punishment will follow detection. Detection and monitoring systems are needed to enforce accountability to collect evidence on corrupt practices. However, these systems

were reported to be generally inadequate in Vietnam's public healthcare sector.

6. *Pressures for Corruption*: The pressures on companies to offer inducements included pharmaceutical market-related and product-related factors. Medicine's low quality, efficacy, and reputation also pressured pharmaceutical companies to resort to inducements. For example, doctors are easily persuaded to prescribe innovator brands and generic medicines from Europe as these were believed to be highly reputable, of high quality, and efficacy. By contrast, Asian generic medicines traded by domestic companies were perceived to be lower in quality and efficacy. Therefore, companies resorted to strategies such as *buyouts* to create incentives for prescribers. With this strategy, *all the needs of a prescriber, such as a car, money, and so forth, were satisfied*. In return, the prescriber had to work for these pharmaceutical companies.
7. *Remuneration Systems and Financial Pressure*: The most frequently reported factor leading prescribers to engage in corrupt practices is inadequate remuneration. Some participants said their (doctors') salary was ranked 17th among 19 professions paid directly by the government. For example, the salary of a public doctor with 20 years of experience was reported to be less than the salary of an engineer who had just graduated and was working in an IT, or banking industry.

Study participants believed low wages led some prescribers to take up inducements to augment their income. For example, an internal medicine doctor participant admitted that the main income of doctors in internal medicine is from pharmaceutical companies by way of commission.

**Ethical Behavior**

Medical ethics weakened when the financial gain from corrupt practices *increased to a level that was rewarding high enough*.

There was also confusion over what was deemed to be ethical behavior. For example, some doctor participants in the study considered it ethically acceptable to take offers from pharmaceutical companies, such as precious gifts, holiday packages, and even commissions, as long as the offer was given without asking for it or after they recommended a particular medicine. For example, in a group interview, an internal medicine doctor shared that *they (the pharmaceutical companies) come and give me cash every month. They said, "You have used this quantity (of products), so I give you this amount of money." However, I don't know the commission accounting for how many percent of the sale, and I don't know if I have prescribed many or few, and I don't care about this. I only know that I have selected their medicine, and they come to thank me*.

**Physician Payments by Pharma: When Did They Begin?**

The study participants pointed out that there was virtually no competition among pharmaceutical companies before the health sector reforms in 1989 when only some State-owned pharmaceutical enterprises produced and supplied medicines to the domestic market. Doctors did not have that behavior of taking commissions from pharmaceutical companies.

The health sector reforms introduced foreign and domestic private companies to the market, creating significant competition. With increasing competition, companies seemed to resort to additional strategies, such as using a sale-based compensation policy to motivate sales staff and paying key opinion leaders, who were chiefs and played a decisive role in the procurement and use of medicines in the hospital. In addition, according to doctor participants, key opinion leaders were given *gifts by way of stocks*, becoming significant shareholders of the companies.

It is only a strict monitoring and enforcement system that can ensure accountability can stop the corrupt and unethical practices in the pharmaceutical industry. A head of a surgical department in an individual in-depth interview metaphorically explained. He said:

*The healthcare sector in Vietnam is like a tiger. When the tiger is not hungry, it doesn't have any need. People are the same. When they don't have to worry about money, they don't have a need to earn money. To do so, we need an appropriate direct remuneration system.*

On the other hand, we need a functioning monitoring system. Although the tiger is full, we still need an electric stick and iron fence; otherwise, it can jump out anytime. People are similar, human beings are good in principle, but when they have the opportunity, they are easily corruptible in the absence of penalties.

The pharmaceutical companies association has updated its ethical code, which The Vietnam Pharmaceutical Companies Association Code (VNPCAC) 2022, is a voluntary code  that defines global practice guidelines for physicians and guidelines for advertising and promoting prescription drugs that is more or less based on IFPMA (International Federation of Pharmaceutical Manufacturers Association) code of ethical conduct.

(Source: Adapted from a research paper by Tuan A. Nguyen, Rosemary Knight, Andrea Mant, Husna Razee, Geoffrey Brooks, Thu H. Dang, and Elizabeth E. Roughead, *Corruption Practices in Drug Prescribing in Vietnam* Nguyen et al. BMC Health Services Research (2018) 18:587 https://doi.org/10.1186/s12913-018-3384-3;(http://creativecommons.org/licenses/by/4.0/),

CASE

# 9

# Unethical Pharmaceutical Marketing Practices in India!

The rampant unethical and corrupt practices in pharmaceutical marketing in India have created a perception that pharmaceutical companies have a vice-like grip over many doctors.

How did the industry that contributed to saving people's lives come to such a stage losing its reputation from a life-saving industry to the one that puts profits before the patients it is supposed to serve?

When did the fall on the reputation ladder begin? The Indian pharmaceutical industry in 1947, at the time of India's independence, was dominated by multinational pharma companies and mainly relied on imports. The country also recognized pharmaceutical product patents till 1970. There were few pharmaceutical companies, and no unethical practices were known then.

In 1970, the Indian government changed its patent laws and recognized only process patents, not product patents, in the pharmaceutical sector. This was a tipping point for the Indian pharmaceutical industry, contributing significantly to pharmaceutical raw material manufacturing development and achieving a leadership position in the generics drugs market globally today. In addition, this change in patent legislation encouraged many entrepreneurs to enter the pharmaceutical market in India with their generic formulations, and the number of companies multiplied quickly.

Fourteen years later, legislation made in the US in 1984 to speed up the introduction of off-patented drugs in the US to contain the increasing healthcare costs positively impacted the growth of the generics and branded-generics industry in India. The 1984 Waxman-Hatch Bill increased the introduction of generic versions of off-patented drugs in the US, and helped the growing Indian branded-generics industry. As a result, the number of companies grew, and about thirty pharmaceutical companies actively and aggressively pursued their goal of entering the lucrative generics market in the US and Europe. So far, these companies have been focusing on international markets in South East Asia, Africa, the Middle East, and Latin America that were relatively less regulated than the highly industrialized and regulated markets, such as the United States, Western Europe, and Japan. Furthermore, when India became a signatory to the GATT (General Agreement on Tariffs and Trade) in 1994 and joined the WTO (World Trade Organization), the race to these international markets and competition in the domestic market became more intense.

Suddenly there was a proliferation of generic versions of new products patented elsewhere worldwide, as product patents were not recognized in India until 2005. Therefore, it is reasonable to presume that unethical practices as we know them now began in the mid-1990s when competition intensified.

Dr. Arun Gadre and Dr. Abhay Shukla, social activists associated with an NGO (Non-Governmental Organization) SATHI (Support for Advisory and training to Health Initiatives), wrote *Dissenting Diagnosis: Voices of Conscience from the Medical Profession*, which presents the magnitude of corrupt and unethical practices between the pharmaceutical industry and physicians today. The authors have interviewed 78 rational-minded doctors across multiple specialties spread over the entire country, who want to stop these unethical practices. In addition, these doctors candidly express their observations and experiences about the current pharma-physician engagement situation. The following comments and observations vividly paint a picture of the unethical and corrupt practices prevailing in the Indian healthcare market.

- For example, the local branch of the doctors association in a large metro city in India decided by secret ballot that henceforth, all their CME (Continuing Medical Education) workshops would not be conducted through subscription from the members but would be sponsored by pharmaceutical companies. As a result, the subjects covered would not be decided by the members but by the pharmaceutical company sponsoring them.
- A practicing physician in Pune, a city in western India, stated that pharmaceutical companies take doctors on foreign trips. They make all arrangements. And there is just a pretense of doing some study on these trips. But, unfortunately, many doctors enjoy all this.
- Today even many senior doctors prescribe eight to ten medicines when a few tablets suffice. There is no doubt that pharmaceutical companies have promoted this practice.
- An ophthalmologist from a medium-sized city in India said: *Some of my doctor friends boast they have traveled the world, sponsored by pharmaceutical companies. One was telling me with pride that even their shirts, pants, vests, and underwear are given by pharmaceutical companies*.
- A big city surgeon remarks: *Pharmaceutical companies sponsor conferences where nobody bothers to listen to lectures. Doctors go to the stalls and collect gifts. They enjoy the free drinks*.
- A pediatrician from a big city mentioned: *Our branch (of a doctors' association) was functioning well. We would organize CME workshops with our funds. Gradually, the pharmaceutical companies pushed their way in. From 1995 onwards, they began to organize their own CME workshops. Earlier, we would focus on the issues of importance that we had decided upon. But then, the pharmaceutical companies began to select only those*

*topics that would help them promote their new drugs. The workshops were free, with liquor thrown in. Finally, the doctors in our city decided that all workshops, from now on, would be organized by pharmaceutical companies. I would ask them why they couldn't spend ₹ 1,000 ($12.5) annually on their education. Why do you want it free? Finally, through a secret ballot, my opposition was set aside, and the basic principles of our (Doctors' association) branch were changed in favor of the pharma companies. I withdrew from it. Now all workshops in our city are conducted by pharmaceutical companies.*

- A pediatrician from Delhi city stated: *A rampant malpractice is in prescribing vaccines— it is organized and takes place on a large scale in a planned fashion. The practitioner gets a cut on the MRP (Maximum Retail Price). The more expensive the vaccine, the higher the cut. The cut is even more than the consultation fee. The doctor gets both— the cut from the company and the fee from the patient.*
- Another specialist from Delhi shared: *The pharmaceutical companies are like a pack of wolves. They keep bugging you and encouraging you to accept some incentives. But, once you take anything from them, they immediately become arrogant. Now they begin to dictate terms, and because you have accepted money and gifts, you are morally bound to them.*
- A dermatologist from a big city said: *The pharmaceutical companies have created mayhem. Nowadays, doctors take money from pharmaceutical companies and prescribe ten to twenty medicines in a single prescription. For example, an antioxidant tablet is always prescribed, whether it serves any purpose or not.*
- *Another specialist from a large city remarked: For example, they don't even produce Doxycycline, an established antibiotic capsule, which costs less than one rupee per capsule. Instead, they add a useless component like lactobacillus with Doxycycline and then sell each of those capsules at a price that is five times*

*more. And when the ordinary or plain Doxycycline capsule is not made available by design, one has no choice but to prescribe the expensive alternative. This is happening without any check concerning many medications.*

These are just a few examples that indicate the level of unethical and corrupt practices in India's pharmaceutical industry and healthcare.

**Healers or Predators?**

In 2020, Dr. Samiran Nundy, emeritus consultant in gastroenterology at Sri Gangaram Hospital in Delhi, Kesav Desiraju, former Union health secretary, and Sanjay Nagral, a Mumbai-based surgeon edited a book— *Healers, or Predators? The Healthcare Corruption in India*. Amartya Sen, the Nobel Laurette, wrote the foreword to the insightful book, which is a collection of 41 investigative essays by doctors practicing individually and with hospitals in India and abroad, public health academics, researchers, activists, civil society members, and a journalist.

Dr. M. K. Mani, a renowned nephrologist from Chennai, India, wrote an essay on Corruption in *Everyday Medical Practice* in the book. The following excerpts from his essay present a clear picture of the prevailing corrupt practices:

- Diagnostic laboratories reward doctors who send patients to them for tests and procedures. They try to legitimize these cutbacks by giving them the label of *interpretation charges* or sometimes calling it *fees for assistance with a procedure*. Practicing doctors are only too happy to channel their patients to these centers, often for unnecessary investigations or procedures.
- A group of patients made an effort to stop this practice themselves. They printed and distributed a public notice which stated that 90 percent of Chennai doctors cheat their patients by accepting commissions from CT, MRI, and Ultrasound scan centers. The pamphlet asked: *Whose money is this,* and went

on to answer: *The Patient*. The pamphlet named six centers and suggested that the public avoid them and preferably go to public hospitals for scans if their doctor asked for one.

- The pharmaceutical industry has been a willing partner in our (doctors') wealth. You will often find doctors prescribing expensive new antibiotics when much cheaper, and older alternatives would do just as well. This is usually because the industry offers inducements— in the form of entertainment, funding overseas trips, or expensive gifts— to doctors who prescribe high-cost remedies. Companies often keep track of the prescriptions of practitioners and keep account of the money brought in by them.
- In 2011, the *Citizen, Consumer, and Civic Action Group*, a Chennai -based NGO formed several years ago to protect citizens' rights in several areas, organized a national seminar on the Regulation of Promotional Practices by Pharmaceutical Companies and presented the findings of a survey they had conducted of medical practitioners, pharmaceutical representatives, pharmacies, hospitals and members of the public. From the findings, it was clear that many doctors are influenced to prescribe certain drugs by the inducements offered by the pharmaceutical industry. Pharmaceutical companies kept track of the prescriptions made by doctors, and the rewards offered to the doctors were proportional to the sales of their drugs.

**UCPMP: A Paper Tiger?**

In 2011, the Department of Pharmaceuticals (DOP)developed a voluntary code known as the Uniform Code for Pharmaceutical Marketing Practices (UCPMP) to arrest the growing unethical marketing practices of bribing doctors by pharma companies. Still, it had no effect as the unethical practices continued unabated. So later, in 2015, the DOP came out with a revised UCPMP, which was also voluntary. However, as the code lacked penal provisions to deter the wrongdoers, many pharma companies continued bribing doctors. Therefore, in September 2015, the government devised a plan to make the marketing code mandatory with legal backing and penal provisions by introducing it under the Essential Commodities Act 1955. But, due to stiff opposition from the industry, the issue is still entangled in bureaucratic red-tapism. As a result, even after more than a decade since its original proposal, UCPMP remains a paper tiger.

**The PIL**

Finally, in 2022, the Federation of Medical and Sales Representatives Association of India (FMRAI) filed a PIL (Public Interest Litigation), asking the government to make the UCPMP a statutory code.

The PIL shows that over eight years, the top seven pharma companies spent ₹ 34,187 crores ($4.27 billion) in marketing. The data cited in the PIL shows that these costs make up 20 percent of the cost of drugs and include direct and indirect benefits to doctors, such as gifts, entertainment, trips, and hospitality.

The PIL, therefore, seeks to make pharma companies criminally liable for bribing doctors through freebies, as in this situation, the customer pays for branded medicines that are *over-prescribed* or *irrationally* prescribed by doctors as per the PIL.

While the jury is still out, the unethical marketing practices continue.

(Source: 1. Dr. Arun Gadre, and Dr. Abhay Shukla, Dissenting Diagnosis: Voices of Conscience from the Medical Profession, Random House India, 2016, 2. Edited by Samiran Nundy, Keshav Desiraju, Sanjay Nagral, Healers, or Predators? Healthcare Corruption in India, Oxford University Press, 2018, 3.Viveka Roychowdhury, Will FMRAI's PIL Finally Prod Govt. to Make UCPMP Mandatory? ExpressPharma, September 1, 2022)

CHAPTER

4

# Transactional Marketing

I want to do business with a company that treats emailing me as a privilege, not a transaction.

*— Andrea Mignolo, Executive Coach, Designer, Seeker*

## Transactional Marketing in Pharma

### What is Transactional Marketing?

Transactional marketing in the pharmaceutical industry is about selling a specific product or service to a target audience. Therefore, it is focused on short-term sales goals and often involves tactics such as direct-to-consumer advertising, sales promotions, and discounts.

Transactional marketing in the pharmaceutical industry often involves promoting specific medications to patients and healthcare providers. This may include direct-to-consumer advertising on television, radio, and print media and targeted advertising to healthcare providers through medical journals and professional meetings.

Pharmaceutical companies may also use sales promotions and discounts to encourage patients to try their products. For example, a company may offer a discount on a new medication for a limited time or provide a free trial to a certain number of patients.

Transactional marketing can also include providing educational materials and resources for healthcare providers to help them better understand the benefits and risks of a particular medication.

The transactional marketing approach has been criticized for being too focused on short-term sales goals and neglecting long-term public health goals. Some argue that it can be misleading, lead to over-diagnosis and over-treatment, and contribute to rising healthcare costs. Transactional marketing can be successful in the short term, but the pharmaceutical industry must also focus on building trust, transparency, and ethical practices.

### Side Effects of Transactional Marketing

Transactional marketing in the pharmaceutical industry has been criticized for potentially contributing to corrupt practices. Focusing on short-term sales goals can pressure companies to engage in unethical behavior to boost profits.

One example of this is the practice of *ever-greening,* which refers to making slight changes to a medication to extend its patent protection and prevent competition from generic drugs. This can result in higher patient prices and increased profits for the company.

Transactional marketing can also lead to over-diagnosis and over-treatment as companies may promote their products to treat conditions that are not well-understood or that may not require medication. This can lead to unnecessary harm and increased healthcare costs.

Another side effect of transactional marketing in pharma is the potential for false or misleading claims about a drug's safety or efficacy. Pharmaceutical companies may overstate the benefits of their drugs or downplay potential side effects to increase sales. This can lead to individuals receiving treatments that may not be appropriate for their condition or that may cause harm.

Additionally, some companies have been accused of hiding negative clinical trial results and selectively publishing positive results to promote their products. This can lead to patients being prescribed medications that are not safe or effective.

Furthermore, transactional marketing can also promote drugs to vulnerable populations, such as children, or target marketing efforts to individuals at risk of overusing or misusing the drugs. Again, this can lead to harm and potential addiction in these populations like the use of OxyContin and similar prescription drugs did.

Also, transactional marketing can negatively impact healthcare costs. Promoting the use of expensive drugs when cheaper alternatives may be available can lead to higher healthcare costs and reduced access to necessary treatments for individuals without adequate insurance coverage.

A far more serious side effect of transactional marketing is that many pharmaceutical companies have been accused of bribing doctors and providing kickbacks to pharmacy benefits managers to boost sales.

It's important to note that not all companies engage in corrupt practices, and many are committed to ethical behavior. However, the pressure to increase sales and profits can lead some companies to engage in unethical behavior.

Despite the legislation in several countries to curb corruption, the global pharmaceutical industry is rife with bribery. Bribery means giving, offering, or receiving an improper benefit to influence the behavior of someone to obtain or retain a commercial advantage. Bribery can take various forms, such as offering, giving money, or anything else of value. Even common business practices or social activities, such as hospitality, can constitute bribes in some circumstances.

Pharmaceutical marketing primarily depends on building close relationships with HCPs (Healthcare Professionals), such as doctors who prescribe their products to patients. The relationship between doctors and the pharmaceutical industry is important as it can help advance drug research, provide continued medical education (CME), and post-marketing surveillance of newly discovered drugs. However, the industry is under the pressure of recouping R&D costs and maximizing profits which are becoming increasingly difficult due to intensifying competition, increasing genericization, and continuing cost containment pressures from the governments and payers. As a result, many pharmaceutical companies seem to explore ways and means of unethical pharmaceutical marketing approaches to maximize profits.

## Pharma-Physician Nexus

The relationship between pharmaceutical companies and physicians has been a topic of much discussion and debate in recent years. While whether a marriage of convenience or made in heaven is debatable, that there is a natural affinity between doctors and the companies that make drugs is beyond debate. Pharmaceutical products are a major part of modern medicine, without which doctors would be far less able to help people. The drug firms, conversely, would not be able to earn revenues from the products they produce without the collaboration of doctors who prescribe them.

When you explore the global landscape of the pharma-physician nexus and examine the various factors that influence this relationship, one key factor that shapes the pharma-physician nexus is the role of financial incentives.

Pharmaceutical companies often provide financial support to physicians through research grants, speaker fees, and other forms of compensation, which can help physicians to conduct research, attend conferences, and stay up-to-date with the latest developments in their field. However, it can also lead to conflicts of interest and potentially compromise the integrity of medical practice, which is the point of major debate.

Another important factor that affects the pharma-physician nexus is the regulatory environment. In many countries, strict rules and guidelines are in place to ensure that the relationship between pharmaceutical companies and physicians is conducted transparently and ethically. For example, the United States has implemented the Physician Payments Sunshine Act, which requires pharmaceutical companies to disclose any payments of gifts they provide to physicians. Similarly, the European Union has implemented the EU Transparency Directive, which aims to promote transparency in the relationship between pharmaceutical companies and healthcare professionals.

The pharma-physician nexus also plays a critical role in the education and training of medical professionals and the development of new drugs and treatments. Many pharmaceutical companies provide educational materials, training programs, and other resources to help physicians stay up-to-date with the latest development in their field. However, there are concerns that these materials may be biased or influenced by the interests of pharmaceutical companies.

Therefore, it is important to ensure that this relationship between doctors and pharmaceutical companies is conducted transparently and ethically, always putting the interests of patients first.

## Bribes and Kickbacks

What is bribery? Saswati Soumya Sahu explains bribery clearly in her article on *Anti Bribery Legislations' Impact on the Pharmaceutical industry* in the SecJure blog published on June 28, 2020. Bribery is a form of corruption and consists of authorizing a bribe, giving or offering a bribe, agreeing to give or offer a bribe, requesting, demanding, or accepting a bribe, or offering or agreeing to accept a bribe. Further, to 'bribe' means to directly or indirectly (through third parties) give, offer or agree to give or offer a loan, reward, advantage, or benefit during the business. It could be cash, a loan, gifts, excessive hospitality or entertainment, or anything else of value. Bribery takes many forms, such as kickbacks, facilitation being more common.

A kickback is also a form of bribery or negotiated bribery, in which an agreed-upon commission or payment is paid to the bribe taker in exchange for services rendered, such as ensuring that a particular contract is awarded to the firm that pays the kickback.

A facilitation payment is another form of bribery. These are relatively small, unofficial payments that companies pay in exchange for providing or expediting routine, non-discretionary government or other services to which a person or company is legally entitled without making such payments.

Here is a list of what a bribe could include:

- Cash and cash equivalents (gift cards or gift certificates, and gold coins)
- Gifts, entertainment, and hospitality
- Payment of travel expenses—especially when there is not a clear business purpose for the trip
- Paying government official to ignore an applicable customs requirement or to accelerate a tax refund
- Vacations
- Offers of employment or other benefits to a family member or friend of the individual

- Political party and candidate contributions
- Charitable donations and sponsorships
- In-kind contributions, investment opportunities, positions in joint ventures, and favorable or steered subcontracts
- Gifts or contributions benefit the government official directly or another person, such as a family member, friend, or business associate.

### *Pay-to-Prescribe* Bribes

Doctors in a nationalized healthcare system and independent practices usually have greater discretion in deciding which drugs to prescribe, allowing pharmaceutical companies to bribe these doctors to prescribe their prescription drug brands more often than necessary. These kickback bribery payments, also known as pay-to-prescribe bribes, are a very common source of corruption in the industry, particularly in branded-generics markets, such as India. Moreover, these pay-to-prescribe bribes are perhaps more insidious than bribes aimed at cheating regulatory approval, as they often take the form of more discreet, non-monetary gifts, such as luxurious, all-expense paid travel under the guise of an educational opportunity or reward programs for high-prescribing doctors. Further, pay-to-prescribe bribes are more difficult for enforcement agencies to detect due to their more subtle nature.

### Bribing Foreign Officials to Get Approvals

Another major source of bribing in pharma is bribing foreign officials in international markets to get marketing approval for their drugs. The global pharmaceutical industry, by nature, requires more approvals in any country for marketing its products than other industries, requiring contacts with foreign officials. This frequent and close contact with foreign officials increases the temptation to bribe, as companies seek to achieve a more efficient and favorable regulatory approval process by *buying off* officials.

Since getting regulatory approvals in domestic and international markets is crucial for the survival and growth of drug companies, some pharma companies have proven that they are willing to do almost anything to gain regulatory approval, and bribery, also known as facilitation or grease payments, is the vehicle used often to achieve this.

In 1977, the US government enacted FCPA (Foreign Corrupt Practices Act), one of the most important corporate statutes in the world, to ban companies from bribing foreign officials to win business. In addition, it requires publicly traded companies to keep books and records that adequately reflect financial transactions.

### Anti-Bribery Legislation

The rest of the world followed suit, and in 1997, 33 countries signed the agreement of the Anti-bribery Convention in Paris. In 2005, 140 countries signed the United Nations Convention Against Corruption (UNCAC), and by 2021, 189 countries became signatories to the UNCAC.

However, despite anti-bribery legislation, bribery remains a very serious, ongoing issue within Big Pharma and, indeed, the pharmaceutical industry. The following case studies illustrate this.

CASE

10

# FCPA Enforces Action Against Pfizer-Wyeth!

In 2012, Pfizer, a global pharma major, agreed to pay $60 million to settle charges alleging that some of its foreign subsidiaries bribed doctors and healthcare officials to gain regulatory approval for the company's drugs and boost sales in those countries.

The Securities and Exchange Commission (SEC) and the Department of Justice (DOJ) negotiated two civil settlements — one with Pfizer and another with Wyeth (acquired by Pfizer in 2009)— for a total of $45 million. The settlement resolves charges of misconduct in about a dozen countries, including Bulgaria, Croatia, Kazakhstan, and Russia. In addition, Pfizer agreed to pay another $15 million in penalties under the Justice Department deal after admitting that the company paid $2 million in bribes to foreign government officials and reaped $7 million in profits.

Kara Brockmeyer, who heads the SEC (Securities and Exchange Commission) unit that enforces the Foreign Corrupt Practices Act, commented that Pfizer subsidiaries in several countries had bribery so entwined in their sales culture that they offered points and bonus programs to reward officials who proved to be their customers improperly.

In its complaint, SEC alleged that Pfizer's misconduct dates to 2001 and detailed some violations in a country-by-country breakdown.

For example, the company's China subsidiary created programs that enabled government doctors to accumulate points based on the number of Pfizer prescriptions they had written and then redeem them for medical books, cellphones, and other gifts.

Further, the SEC complaint added that Pfizer's China employees also used travel incentives to bribe doctors. For example, in 2006, a Pfizer China marketing manager told his regional sales manager that the company would foot the travel bill for two doctors attending a conference in Australia only if the doctors promised to use no fewer than 4,200 injections a year and to prescribe a Pfizer product to more than 80 percent of their patients.

It is worth noting that no one at Pfizer headquarters knew of the bribery. The company even flagged regulators for improper payments in Croatia as soon as they knew about it in 2004, cooperating with federal investigators.

Amy Schulman, executive vice president and general counsel for Pfizer, stated in this context: *The actions which led to this resolution were disappointing, but the openness and speed with which Pfizer voluntarily disclosed and addressed them reflects our true culture and the real value we place on integrity and meeting commitments*.

(Source: Article written by Dina ElBoghdady and published in The Washington Post on August 7, 2012)

CASE

11

# SEC Charges Eli Lilly and Company with FCPA Violations!

The Securities and Exchange Commission (SEC) charged Eli Lilly and Company on December 20, 2012, with violations of the Foreign Corrupt Practices Act (FCPA) for improper payments its subsidiaries made to foreign government officials to win millions of dollars of business in Russia, Brazil, China, and Poland. The allegations are:

- Lilly's subsidiary in Russia paid millions of dollars to offshore entities for alleged *marketing services* to induce pharmaceutical distributors and government entities to purchase Lilly's drugs, including approximately $2 million to an offshore entity owned by a government official and approximately $5.2 million to offshore entities owned by a person closely associated with an important member of Russia's Parliament. Despite the company's recognition that the marketing agreements were being used to create sales potential with government customers and that it did not appear that any actual services were being rendered under the agreements. Eli Lilly allowed its subsidiary to continue using the agreement for years.
- Lilly's subsidiary in Brazil allowed one of its pharmaceutical distributors to pay bribes to government health officials to facilitate $1.2 million in Lilly's drug products sales to state government institutions.

- Lilly's subsidiary in Poland made eight improper payments totaling $39,000 to a small charitable foundation founded and administered by the head of one of the regional government health authorities in exchange for the official's support for placing Lilly's drugs on the government reimbursement list.

Further, the SEC alleged that when the company did become aware of possible FCPA violations in Russia, Lilly did not curtail the subsidiary's use of the marketing agreements for more than five years. Antonia Chion, Associate Director of the SEC Enforcement Division, said, "*we strongly caution company officials from averting their eyes from what they do not wish to see.* "

Kara Novaco Brockmeyer, Chief of the SEC Enforcement Division's Foreign Corrupt Practices Unit, added that *Eli Lilly and its subsidiaries possessed a check-the-box mentality regarding third-party due diligence. Companies can't simply rely on paper-thin assurances from employees, distributors, or customers. Instead, they need to look at the surrounding circumstances of any payments to adequately assess whether they could wind up in a government official's pocket.*

Lilly consented to the entry of a final judgment permanently enjoining the company from violating the FCPA's anti-bribery, books and records, and internal controls provisions without admitting or denying the allegations.

(Source: US Securities and Exchange Commission's Press Release dated December 20, 2012)

CASE

12

# Bristol-Myers Squibb Pays $14 Million to Resolve China FCPA Offenses!

The Securities and Exchange Commission (SEC) said on October 5, 2015, that Bristol-Myers Squibb (BMS) agreed to settle charges that its joint venture in China made cash payments and provided other benefits to healthcare providers at state-owned and state-controlled hospitals in exchange for its prescription drug sales.

Between 2009 and 2014, BMS China sales representatives tried to win and increase business by giving healthcare providers cash, jewelry, other gifts, meals, travel, entertainment, and sponsorships for conferences and meetings.

According to the SEC, BMS china inaccurately recorded the spending as legitimate business expenses in its books and records of Bristol-Myers Squibb.

The SEC found the following FCPA (Foreign Corrupt Practices Act) violations by Bristol-Myers Squibb and included them in its order:

- Bristol-Myers Squibb failed to respond effectively to red flags indicating that sales personnel provided bribes and other benefits to generate sales from healthcare providers in China.
- Bristol-Myers Squibb did not investigate claims by certain terminated employees of China that faked invoices, receipts, and purchase orders were widely used to fund improper payments to healthcare providers.

- Bristol-Myers Squibb was slow to remediate gaps in internal controls over interactions with healthcare providers that were identified repeatedly in annual internal audits of BMS China between 2009 and 2013.

The SEC resolved the enforcement action through an internal administrative order and did not go to court. Bristol-Myers Squibb consented to the order without admitting or denying the findings and disgorged to the SEC $11.4 million of profits plus prejudgment interest of $500,000 and paid a civil penalty of $2.75 million. In addition, the company agreed to report to the SEC for two years on the status of its recommendation and implementation of FCPA and anti-corruption compliance measures.

Kara Brockmeyer, chief of the SEC's FCPA enforcement unit, said: *Bristol-Myers Squibb's failure to institute an effective internal controls system and to respond promptly to indications of significant compliance gaps at its Chinese joint venture enabled a widespread practice of providing corrupt inducements in exchange for prescription sales to continue for years*.

(Source: The FCPA Blog article by Richard L. Cassin published on October 5, 2015)

CASE

13

# GlaxoSmithKline's Bribery Scandal in China!

Since China opened up its economy to international companies decades ago to help develop it, Multinational companies have turned to China as it offered a major growth opportunity with a large population and showing major economic growth. As a result, many multinational companies, including pharmaceutical companies, avoided scrutiny over bribery as their presence helped establish manufacturing in China and created jobs. However, this policy has been going easy on foreign enterprises for a long time.

Benjamin Shobert, an expert contributing writer to Forbes on healthcare and aging in China, suggests *Three Ways To Best Understand GSK Scandal in China* in his article with the same title published on Forbes.com on September 3, 2014.

Firstly, in 2012, Chinese President, Xi Jinping, began a crackdown on corruption and started a reform process in China to make the country globally competitive, including in pharmaceuticals. The GSK scandal, along with other multinational pharmaceutical companies, was a part of the growing anti-corruption initiatives and was a high-profile example.

Secondly, Xi Jinping, in addition to a crackdown on corrupt practices prevalent in the Chinese business environment, also initiated a massive, once-in-a-generation expansion of its healthcare system.

Extensive bribery by pharmaceutical multinationals is also a part of China's massive expansion of the healthcare system. China has built new hospitals and healthcare infrastructure at a rapid pace. However, according to Benjamin Shobert, Senior Associate for International Health at the National Bureau of Asian Research, doctors in these hospitals are often overworked and underpaid, and hospital administrators could not close the gap between government reimbursements and the growing costs of healthcare. As a result, hospital administrators and doctors have found alternative means to make up for the revenue not provided by the governments. Administrators have incentivized doctors to prescribe unnecessary medicines, surgical products, and diagnostic evaluations. Doctors have supplemented their inadequate incomes through bribes from pharmaceutical companies. Further, patients' families made red envelope payments directly to doctors to ensure proper and timely care. These practices created problematic inefficiencies within China's healthcare system, which has been trying to improve health benefits for the Chinese people.

Initially, when China opened up its economy to international companies decades ago to develop its economy, the country had followed a policy of going easy on foreign enterprises, according to Jerome Cohen, a legal advisor for international companies in China. He added that the government didn't want to cause embarrassment or give outsiders the impression that China is plagued with corruption. That changed in the later years and American and European companies with significant investments in China observed that the business environment was less hospitable than they thought. Many businesses believed China's various ministries held them to higher regulatory standards than their Chinese competitors.

Multinationals such as GSK had to navigate the complex field where international compliance standards overlap with how healthcare is consumed and paid for in developing healthcare economies, such as China because the country's leadership believed that in the absence of strong leadership positions by the domestic companies, the country's overall economic growth would be difficult to achieve.

GlaxoSmithKline's bribery scandal in China laid bare the corruption in the healthcare sector in China in 2013.

### GSK's Bribery Scandal

It all began in 2013 with an anonymous whistleblower sending an e-mail to GlaxoSmithKline board members describing fraudulent activities in China. The whistleblower stated in the e-mail that medical professionals were given all-expenses-paid trips under the pretense of attending professional conferences. He also stated that Lamictal (Lamotrigine) was being heavily promoted as a treatment for bipolar disorder, despite being approved only for the treatment of epilepsy by Chinese regulators. The whistleblower explained in his e-mail that GSK almost killed one patient by illegally marketing Lamictal and that GSK China bought the patient's silence for $9,000. The whistleblower sent e-mails to GSK's board members and auditor PriceWaterCoopers over seventeen months.

The Chinese State media reported on the investigation that GSK apparently bribed government officials, gifting a Shanghai investigator an iPad and treating him to a $1,200 dinner. Also, the media reported that Mark Reilly, the GSK country manager for the Chinese market, was given company funds to bribe Beijing officials.

Following the trial in Changsha in September 2014, involving Mark Reilly and four executives of GlaxoSmithKline in China, all were sentenced to prison time, which is in addition to a hefty $490 million fine to the company. Mark Reilly subsequently received a suspended sentence and was also deported from China. The charges against GSK have also affected other companies' presence in China. For example, Microsoft faced scrutiny over antitrust allegations; Apple modified its tax practices after facing fines in China.

GlaxoSmithKline apologized for its dealings in China and pledged to reduce and change its interactions with Chinese healthcare professionals. Andrew Witty, the Chief Executive Officer of GSK, stated: *We will also continue to invest directly in the country to support the government's healthcare reform agenda and long-term plans for economic growth*.

(Source: Adapted from articles—(1) Benjamin Shobert, Three Ways To Understand GSK's China Scandal, forbes.com, September 4, 2013, (2) Ethics Unwrapped, Curbing Corruption: GlaxoSmithKline in China, McCombs School of Business, (3) GSK China scandal. (2022, August 4). In *Wikipedia*. https://en.wikipedia.org/wiki/GSK_China_scandal)

CASE 14

# Teva Pharmaceuticals Pays Nearly $520 Million, the Largest Fine to US Criminal and Regulatory Authorities for the FCPA Violations!

Teva Pharmaceuticals, the world's largest manufacturer of generic pharmaceutical products, and its wholly owned Russian subsidiary, Teva Russia, agreed to resolve criminal charges for violating the FCPA and pay a criminal penalty of nearly $520 million to US criminal and regulatory authorities.

The company admitted that Teva executives and Teva Russia employees paid bribes to a high-ranking Russian government official intending to influence the official to use his authority to increase sales of Teva's multiple sclerosis drug, Copaxone, in annual purchase auctions held by the Russian Ministry of Health. The corrupt arrangement occurred while the Russian government sought to reduce the amount spent on costly foreign pharmaceutical products, such as Copaxone. Between 2010 and 2012, Teva made a repackaging agreement with a company owned by a Russian government official and earned more than $200 million in profits on Copaxone sales to the Russian government through this arrangement. Furthermore, the Russian government official earned approximately $65 million in corrupt profits through inflated profit margins granted to the official's company.

Teva also admitted to paying bribes to a senior government official within the Ukrainian Ministry of Health to influence the Ukrainian government's approval of Teva drug registrations, which were necessary for the company to market and sell its products in the country. Between 2001 and 2011, Teva engaged the official as the company's registration consultant, paid him a monthly fee, and provided him with travel and other things of value totaling approximately $200,000. In exchange, the official used his official position and influence in Ukraine, for Teva pharmaceuticals' products, including Copaxone and insulins.

In addition, Teva admitted that it failed to implement an adequate system of internal accounting controls and failed to enforce the controls it had in place at its Mexican subsidiary, which allowed bribes to be paid by the subsidiary to doctors employed by the Mexican government. Teva admitted that its Mexican subsidiary had been bribing these doctors to prescribe Copaxone since at least 2005. Moreover, Teva executives in Israel responsible for developing the company's anti-corruption compliance program in 2009 had been aware of the bribes paid to government doctors in Mexico. Teva also admitted that its executives put in place to oversee the compliance function were unable or unwilling to enforce the company's anti-corruption practices.

The regulatory authorities, such as Leslie R. Caldwell of the Justice Department, Criminal Division, Stephen Richardson of the FBI's Criminal Investigative Division, and William J. Maddalena of the FBI's Miami Field Office spoke on occasion and here is what each one of them said regarding Teva's FCPA charges and the consequent agreement to pay the huge fine:

Assistant Attorney General Leslie R. Caldwell of the Justice Department's Criminal Division said: *Teva and its subsidiaries paid millions of dollars in bribes to government officials in various countries and intentionally failed to implement a system of internal controls that would prevent bribery. Companies that compete fairly, ethically, and honestly deserve a level playing field, and we will continue prosecuting those who undermine that goal*.

Stephen Richardson, Assistant Director of the FBI's Criminal Investigative Division, said: *No matter where corruption occurs, the*

*FBI and our global partners are committed to diligently rooting out the corruption that betrays the public trust, and threatens a fair economy for all.*

William J. Maddalena, Assistant Special Agent in Charge, FBI Miami, said: *As demonstrated by this case, Teva's egregious attempt to enrich themselves failed, and they will now pay a tough penalty.*

(Source: US Department of Justice Press Release, Teva Pharmaceutical Industries Ltd. Agrees to Pay More Than $283 Million to Resolve Foreign Corrupt Practices Act Charges, December 22, 2016)

To prevent corrupt practices, it is important for pharmaceutical companies to be transparent about their research, development, and pricing practices and to adhere to high ethical standards in all aspects of their operations. The regulatory bodies should also strengthen their oversight and enforcement to ensure that companies are held accountable for their actions.

CHAPTER

5

# Restoring Pharma's Reputation

I have always stuck up for western medicine. You can chew all the celery you want, but without antibiotics, three quarters of us would not be here.

*— Hugh Larie, English Actor, Writer, Musician*

# Restoring Pharma's Reputation

Once considered one of the most reputed industries, the pharmaceutical industry's reputation worldwide has fallen considerably to a level not much better than the financial or tobacco industries. What contributed to this fall? And why is reputation important? Let us explore, and find out whether and how it can be restored.

### Importance of Reputation

A company's reputation is among its most valuable assets. Research confirms this. The percentage of a company's value attributable to tangible assets has dropped from 90 percent to just 25 percent over the past three years, according to the research conducted by the New York-based consultancy firm, Ethisphere. Other estimates too confirm this. They suggest that a company's intangible assets (including reputation) currently represent 40 to 60 percent of its market capitalization.

### Declining Reputation

A recent patient view survey revealed a 19 percent decline in the pharmaceutical industry's reputation in general and its leading companies in particular from the previous year. Furthermore, the research found that only 34 percent of the patients surveyed believed that multinational drug companies have an excellent or good reputation. So what are the reasons for this declining trend in the pharmaceutical industry that has greatly contributed towards the health and well-being of populations worldwide over the years? Here are some of the more important reasons:

1. Pharmaceutical companies have become increasingly marketing-focused, highlighting only favorable clinical trial results that drove product sales. They were hiding unfavorable clinical trial results, and the information was misleading to the physicians as it was neither complete nor accurate. This has corrupted the scientific literature leading to the erosion of trust.

2. In addition, some of the industry's leading pharmaceutical companies have been communicating about the damaging side effects of some of their major products in an evasive manner, encouraging physicians to prescribe their products even in inappropriate clinical conditions.
3. Pharmaceutical companies have been increasing their unethical and illegal marketing practices, leading to increased litigations. Some of the more prominent litigations brought by the law firms involved are Pfizer's Rezulin, GlaxoSmithKline's Paxil, Wyeth's (now Pfizer) Rapamune (Sirolimus), and Merck's Vioxx. As a result of these litigations, lawyers who have targeted asbestos and tobacco manufacturers in the past are increasingly turning their attention to drug companies alleging that they have hidden the harmful side effects of their medicines from consumers.
4. Pharmaceutical companies' corrupt marketing practices, such as paying hospitals and doctors to prescribe, allegations of price-fixing kickbacks, and payments to generic companies to delay generic drug access to the public— all these activities are widely publicized across various media channels.

All these activities and lawsuits have created a public perception that the pharmaceutical industry cares less about its patients than profits. In addition, the following missteps of pharmaceutical companies have reinforced that negative perception.

1. Pharmaceutical companies have failed to assist patients in securing medications in a difficult economic environment.
2. For some time now, pharmaceutical research has been focusing on patentable research and can offer drugs with only short-term benefits.
3. The pharmaceutical industry is not serving the needs of neglected patient groups.
4. Drug companies have been engaging in inappropriate marketing practices.
5. Pharmaceutical pricing policies clearly lack fairness.

6. Pharmaceutical companies talk about patient-centricity these days but do not practice what they preach.
7. Lack of integrity is evident from the industry's corrupt practices and the huge fines they pay for illegal marketing activities such as off-label promotion and suppression of facts regarding adverse effects and unfavorable trial results.

One can observe that much of the decline in the pharmaceutical industry's reputation over the years is self-inflicted. It is possible to undo what it has done. Research suggests that once a company's reputation declines, it would take about 3.5 years to rebuild, even in the best circumstances. In the case of the pharmaceutical industry, it could take even longer, considering the complexities and the number of stakeholders involved. Moreover, it requires both the industry as a whole and individual drug companies to come together and address the factors that precipitated the decline in the first place.

Sherry Baker of Brigham Young University, Utah, US, and David L. Mortinson of Florida International University, Florida, US, proposed a five-part TARES test to define the moral boundaries of persuasive communication. It serves as a set of action-guiding principles directed toward the moral consequence of persuasion. The pharmaceutical industry would do better if it puts its communication to all its stakeholders and the general public through this test as it could help the industry significantly improve transparency restoring its declining reputation.

## The TARES Test

How can the pharmaceutical industry change its behavior to improve its reputation by stamping out unethical marketing practices? Situational factors and organizational culture are the most important determinants of ethical behavior in an organization. In a corporate setting, the influence of one's peers and the top management is even more important than one's personal belief system. Introducing an ethical framework, such as TARES Test may be one way to change behavior and imbibe good marketing practices. While there are many ethical frameworks, TARES Test is simple, self-explanatory, and yet introspective.

The TARES Test provides a useful framework for establishing values and measuring whether or not they are followed specifically for marketing and advertising. Still, the general principles may be applied at an individual and an organizational level. The TARES framework provides a view of ethics both in macro-ethical and micro-ethical contexts. *Micro-ethical context* is looking at a particular piece of persuasive communication or an advertising message as unethical or ethical. How the company as a whole must examine the persuasive communication or an advertising message and its effect on the target audience is *a macro-ethical context.* Marketers must put the interest of their target audiences before their own narrowly defined self-interest, which may be an increase in sales and market share, for example.

The TARES Test combines five principles, hence the acronym—Truthfulness, Authenticity, Respect, Equity, and Social Responsibility. It also suggests the application of the *Golden Rule,* a concept that forms the foundation of equity. The following table presents an illustrative TARES Test questionnaire that helps evaluate an ethical framework and mindset.

**Table 5.1 The TARES Test**

| **Truthfulness** |
|---|
| 1. Is communication true and truthful? |
| 2. Have I presented selected information? |
| 3. Have I withheld information needed to make an informed decision? |
| 4. Have I distorted information by using improper comparisons? |
| **Authenticity** |
| 1. Does this action affect my integrity? |
| 2. Would I recommend this product to my loved ones without reservation? |
| 3. Are my motives self-serving or do they benefit others? |
| 4. Would I be proud if others were to find out my behavior? |
| **Respect** |
| 1. Do my actions respect the rights and dignity of others? |
| 2. Have I provided enough information to meet the needs of physicians and patients? |
| 3. Will physicians and patients benefit from following the advice in my communication? |
| 4. How I can be more responsible in my behavior towards physicians and patients? |

**Table 5.1** *Contd...*

**Equity**
1. Have I targeted vulnerable audiences based on the content of my communication?
2. Are my persuadees (target audience) able to assess the claims I make in my communication fully and rationally?
3. Does the audience know that I am persuading them rather than informing them?
4. Should I engage in this type of persuasion?

**Social Responsibility**
1. Do I respect the welfare of all?
2. Do I harm individuals with my action?
3. Do I encourage public dialogue based on truthful information?
4. How can I serve the interests of the society?

Source: Global Issues in Pharmaceutical Marketing by Lea Prevel Katsanis

## Restoring Pharma's Reputation

Mark Kessel, co-founder and partner at private equity firm, Symphony Capital, suggests the following nine steps pharmaceutical companies must take to restore their reputation in his insightful article, *Restoring the pharmaceutical industry's reputation* in Nature (October 2014). Here is the essence of what he has suggested:

1. *Refocus on Patients' Needs:* The pharmaceutical industry seems to have failed mainly on two counts in living up to its earlier reputation. One is its ethical conduct, and the other is its customer focus. They need to improve the stakeholders' perception that the company cares about patients and is committed to all its customers. The company also needs to demonstrate its ethical conduct through its actions.
2. *Cease DTC Advertising:* Does DTC (Direct-to-consumer) advertising of prescription medicines strengthen patients' perception that drug companies care more about improving their sales to enhance their earnings than they do about improving patients' health? Pharmaceutical companies should examine and reflect on this and desist from DTC advertising.
3. *Follow a Responsible Pricing Strategy and Justify the Pricing:* The cost of developing a new drug and the time it takes to bring a new drug from the concept to the market are very expensive, costing over two billion dollars. The pharmaceutical

industry should educate all its stakeholders about this. In addition, it should justify the pricing of the new drug by explaining the benefits it offers the affected patients and the cost savings it brings to the healthcare system. Above all, a drug company should consider whether its pricing policy should be tempered to avoid the potential outcry from its stakeholders and the long-term impact of its pricing on its reputation.

4. *Restore and Demonstrate an Ethical Culture:* A company's ethical conduct is at the top of the reputation measurement system. In fact, a company's ethical behavior determines its reputation. Suppose companies tolerate unethical behavior of their senior management. In that case, it sends a message or a signal within the organization that it is fine to be unethical as long as you get the results and can circumvent potential liability. Despite adverse media publicity of the pharmaceutical industry's unethical behavior in terms of billions of dollars of fines it paid for its illegal marketing activities, the public does not believe senior management was held accountable. To change this perception, the industry needs to restore ethical behavior, put better controls, and punish misconduct of executives responsible for such misdeeds.
5. *Implement Data Transparency:* There are several allegations that pharmaceutical companies publish successful trial data only, withhold negative data from publication, and even rig study designs to foster favorable outcomes. The only way to erase this negative perception in the minds of the industry stakeholders is to implement data transparency by publishing both positive and negative trial results.
6. *Change Industry Messaging:* Send the Right Messages. Although actions speak louder than words, you must communicate the right things and accentuate the positive. The pharmaceutical industry did not have an effective and consistent program to inform and educate consumers and other stakeholders on how it has contributed to health and improve lives by bringing new life-saving therapies to patients. In its heydays, pharmaceutical companies were run by CEOs

with scientific training and credibility in the marketplace. They were perceived as individuals concerned about patients' health and well-being. Today most pharmaceutical companies are run by lawyers or individuals from sales and marketing (Novartis is an exception), who are not likely to have the same level of respect from stakeholders. In addition, the focus on Wall Street and the quarterly results lead one to believe that the focus is more on stockholders than patients, who are the most important stakeholders for a pharmaceutical company. Pharmaceutical companies need to change this perception. Pharmaceutical companies should encourage their research and development or medical department heads to interact more frequently with stakeholders and explain how their new drugs benefit and why it is worth the price, its development expenses, and the company's assistance programs to improve its access to patients.

7. *Counter the Misinformation:* Certain academicians, politicians, and others seem to claim that academia and the National Institutes of Health in the United States bring new drugs to the market, not the pharmaceutical industry. The pharmaceutical industry, therefore, needs to counter this misinformation. The industry should consider investing in a communication strategy to inform and educate all stakeholders rather than squandering funds on DTC advertising.
8. *Reduce Government Lobbying:* The pharmaceutical industry in the United States has been spending well over US $200 million per year on lobbying activities with the government. The lobbying effort in other countries, too, seems to be considerable. Despite its high lobbying efforts, the pharmaceutical industry in the US has been criticized by Congress and the FDA on several occasions. Moreover, this lobbying effort makes government bodies look at drug companies suspiciously. Therefore, pharmaceutical companies must present the facts with candor and restore confidence in their testimony and data.
9. *Being Good by Doing Good:* Business history is replete with examples where companies that addressed social issues were

the companies that performed well in the marketplace. For the pharmaceutical industry, restoring reputation will mean placing more emphasis on patients and their needs and convincing the shareholders about the importance of patient focus and how it helps improve their value in the long term.

The pharmaceutical industry can and should start working on restoring its reputation immediately if it has not started already. The industry can draw inspiration from what George W. Merck, the legendary CEO of the world's most respected pharmaceutical company during his time, said on December1, 1950 on the occasion of the 50th anniversary of the company at Medical College of Virginia:

*We try to remember that medicine is for the patient. We never try to forget that medicine is for the people. It is not for the profits. The profits follow, and if we remember that, they have never failed to appear. The better we remembered it, the larger they had been.*

**What Pharma Should Do**

While a transactional marketing strategy may appear effective in the short run and cost-effective, it has significant downsides. Consider these, for example:

- One of the biggest downsides to transactional marketing is that it can lead to a decline in a company's reputation. Transactional marketing is often seen as impersonal and focused solely on making a sale rather than building a relationship with the customer.
- Transactional marketing can also lead to a negative reputation when companies are seen to be putting profit over patients' welfare by pushing unnecessary medication or overpricing drugs.

Therefore, changing the transactional marketing practices to patient-centric ones with a long-term view on the horizon, and not on a quarter-by-quarter performance only, is crucial for the pharmaceutical industry to improve and restore its declining reputation to the glorious levels of the past.

Since unethical and corrupt marketing practices are the primary reasons for pharma's declining reputation, stopping those practices would be the first step the industry must take. Pharma should move away from its current *transactional* marketing practices, and to *transformational* marketing.

CHAPTER

6

# Transformational Marketing in Pharma

Engaging in an authentic, meaningful conversation with consumers will be the key to marketing success and growth, even if that means acknowledging negative feedback; Transparency is paramount.

*— Ron Blake, President and CEO, Rewards Network*

## Transformational Marketing in Pharma

Transformational marketing in the pharmaceutical industry is about building long-term relationships with patients and healthcare providers, focusing on creating value for the patients and the health system rather than just promoting a specific product or service. It strives to transform patients' lives through patient education and engagement programs, community outreach and research, and by providing meaningful services above and beyond the pill. Above all, transformational marketing is about ethically conducting its activities and operations.

For example, instead of only promoting a specific drug for a particular condition, a pharmaceutical company may also focus on educating patients and healthcare professionals about that health condition and the various treatment options available.

Companies practicing transformational marketing are truly patient-centric. They engage with patients and patient communities to understand their needs and concerns, building trust by showing commitment to improving public health. More importantly, transformational marketing requires companies to adhere to high ethical standards in all operations, including sales and marketing, interactions with healthcare professionals, and clinical trials, transparently.

# Transactional to Transformational Marketing in Pharma

It is important to understand the two marketing approaches, transactional and transformational marketing before we discuss the need to transform our marketing approach from transactional to transformational.

Transactional marketing is a business strategy focusing on single, *point-of-sale* transactions. It emphasizes maximizing the efficiency and volume of individual sales rather than developing a relationship with the buyer.

In contrast, transformational marketing aims to build strong, long-term relationships between brands and customers, leading to repeat sales and increased customer loyalty. It focuses on enabling authentic, relevant, and transparent communication between customers and companies to build trust and improve reputation.

The following table presents the key differences between transactional and transformational marketing.

**Table 6.1 Transactional vs. Transformational Marketing**

| | Transactional | Transformational |
|---|---|---|
| Objective | Acquire new customers and increase volume of point-of-sale transactions | Improve customer retention and build customer loyalty |
| Length of Relationship | Short term | Long term |
| Customer Contact | Minimal | Frequent |
| Type of Marketing | Mass Marketing and Promotion | Personalized Marketing |
| Type of Promotional Strategy | May include unethical and corrupt marketing practices, with quid pro quo arrangements, such as paying in cash or kind, expensive gifts, luxury holidays and more. | Focus is on building enduring relationships based on mutual trust and respect for each others' competence. |

## Why Transformational Marketing?

While transactional marketing offers some benefits, the potential side effects outweigh the benefits. Moreover, for over twenty years, many pharmaceutical companies seem to be following the practice of transactional marketing literally as a transaction, which is defined as *a finalized agreement between a seller and a buyer for transferring goods, services, or financial assets in exchange for money*.

While in other businesses, that may be an acceptable and appropriate way of marketing, it is against the ethical values and code of pharmaceutical marketing. Many pharmaceutical companies vie with each other to persuade the prescribing physicians to prescribe their products, engaging them in whatever manner it takes, indulging in unethical and corrupt practices, such as paying for prescriptions by cash or kind. In most countries, such practices are unethical, even illegal, and yet companies seem to turn a blind eye to ethical values and principles.

There are several reasons why pharma should immediately shift to transformational marketing from its current transactional practices. Consider these, for example:

1. Firstly, transactional marketing is not sustainable because loyalty cannot be bought but earned. Consider, for example, that your top customers might migrate to your immediate competitor if their offer is right!
2. Secondly, recent technological advancements and the digital revolution have impacted the pharmaceutical industry in many ways. Pharmaceutical companies must adapt to emerging technologies and implement an organization-wide digital transformation program to successfully ride the waves of change.
3. Consumerization of Healthcare is the third major reason that indicates a shift toward a more patient-centric model, where individuals have greater control over their health and healthcare decisions. Therefore, pharma companies should provide patients with tools and resources to help them make more informed decisions, such as cost and quality transparency, empowering them to take a more active role in their care.

4. Outcomes-based pricing and Value-based Agreements. With the growing focus on reducing healthcare costs and improving patient outcomes, pharmaceutical companies must demonstrate the value of products regarding patient outcomes and cost-effectiveness. This will require companies to shift their focus from a volume-based to a value-based model of care.
5. Data-Driven Marketing: With the rise of big data and analytics, pharmaceutical companies increasingly use data-driven marketing techniques to identify and engage with customers more effectively. This includes using predictive analytics to identify the best targets for their marketing campaigns and personalizing messages in real time.
6. Digital and Social Media: With the proliferation of digital technologies and social media platforms, pharmaceutical companies increasingly use digital channels to reach customers and consumers where and when they want to be reached in new and innovative ways. From digital consultations and telemedicine to mobile health apps and wearables, digital technologies enable companies to enhance the patient experience and improve health outcomes.
7. Artificial Intelligence and Machine Learning will play a growing role in the pharmaceutical industry, allowing companies to analyze and process large amounts of patient data, identify the best targets for their marketing campaigns and personalize their messages in real time.
8. The pharmaceutical industry's decision-making power for prescribing drugs is shifting from individual physicians to groups of healthcare professionals and organizations. This is due to several factors, including the increasing use of evidence-based medicine, the rise of accountable care organizations (ACOs), and the growing influence of payers and pharmacy benefit managers (PBMs) in the drug prescribing process. Additionally, regulatory bodies, such as the FDA in the US, are also playing a greater role in decision-making by providing guidelines and recommendations for drug prescribing. This

shift in decision-making power is leading to changes in the way drugs are developed, marketed, and prescribed, with a greater emphasis on efficacy, safety, and cost-effectiveness.

All these reasons suggest that pharma should immediately change its marketing practices from transactional to transformational. In addition, you cannot add the value the customers seek with transactional analysis. Above all, you cannot build trust through transactional marketing.

Therefore, it is no longer whether to shift to transformational marketing, but how?

**Building Trust is the First Step**

Pharmaceutical companies should understand that trust plays a crucial role in transformational marketing.

Transformational marketing aims to create a deeper emotional connection with customers by emphasizing the values and benefits of a product or service. Pharma can achieve this connection only through transparency, consistency, and authenticity because customers are more likely to trust a company that is transparent about its practices, consistent in its messaging and delivery, and authentic in its communication. Trust is also built through positive experiences and customer satisfaction.

Therefore, the key to successful marketing in the pharmaceutical industry is to focus on building relationships with patients and healthcare providers and to create authentic educational and informative content that helps them make informed decisions about their health. This can be achieved through various channels, such as digital marketing, patient engagement programs, and experiential marketing events. Moreover, when customers trust a company, they are more likely to be loyal and recommend it to others.

**How Pharma Can Build Trust**

Pharmaceutical companies can improve their trustworthiness by building trust in their marketing by following these seven strategies:

1. *Transparency*: Be open and honest about drug research, development, and testing. Share information about the clinical trials and the results of the trials.
2. *Compliance*: Follow all regulatory guidelines and laws related to drug promotion and provide accurate and balanced information about the risks and benefits of a drug. Also, follow the ethical code prescribed by the industry association and the government.
3. *Patient-centricity*:

A. Show that the company prioritizes the well-being of patients, highlighting patient testimonials and showcasing the commitment to patient education and support.
B. Put the patients' interests at the center of everything you do.
C. Ensure meaningful patient engagement throughout their journeys.

4. *Authenticity*: Be authentic in your communication. Use real people in advertising and avoid using overly polished or unrealistic images.
5. *Collaboration*: Partner with healthcare professionals, patient organizations, and advocacy groups to build trust and credibility.
6. *CSR*: Demonstrate your company's Corporate Social Responsibility (CSR) initiatives, as this will help to build trust and a good reputation among the public.
7. *Data Privacy*: Implement strict data privacy policies and ensure that patients' data is always protected.

**Creating Vision is the Next Step**

When planning a major change, such as from transactional to transformational marketing, you must have a Change Vision or Vision to Transform, in addition to your Corporate Vision.

A change vision is a specific, short-term vision for an organization's change initiative. In contrast, a corporate vision is a long-term, overarching vision for the entire organization. A change vision outlines the desired outcome of a particular change effort and the steps needed to achieve it. In contrast, a corporate vision defines the overall direction and purpose of the organization and serves as a guide for decision-making and strategic planning. Simply put, a change vision is a small part of the big picture (corporate vision).

### 'Change' Vision

First and foremost, if you are part of an organization trying to drive a major change, from transactional to transformational, you must have a change vision. The vision to transform. It is to present a picture to people of what the organization will look like after they have made significant changes, and it also shows them the opportunities they can take advantage of once they do that. Such a vision motivates people, and it's essential to any successful change you are trying to make.

### 'Corporate' Vision

A Vision and Mission Statement is important for a pharmaceutical company because it defines its overarching goals and objectives and guides decision-making and strategic planning.

A clear vision and mission statement helps to communicate the company's purpose and direction to employees, stakeholders, and customers and help to attract and retain talent. For example, Pfizer's vision is *to be the premier, innovative biopharmaceutical company, discovering, developing, and providing medicines and vaccines that improve people's lives*. This clearly, communicates the company's focus on innovation and commitment to improving health outcomes. Another example is Novartis' mission statement *to discover new ways to improve and extend people's lives* emphasizes the company's focus on research and development and its commitment to improving patient outcomes.

Here are some more examples of the mission statements of some big pharma companies:

- **Roche**: Doing now what patients need next.
- **Merck**: To save and improve lives through innovative medicine.

- **GlaxoSmithKline:** To help people do more, feel better, and live longer.
- **Sanofi**: To be a global biopharmaceutical leader transforming scientific innovation into healthcare solutions.
- **AstraZeneca**: Innovating to deliver life-changing medicines.
- **Bristol-Myers Squibb**: To discover, develop, and deliver innovative medicines that help patients prevail over serious diseases.
- **Johnson & Johnson**: To help improve the quality of human life by providing innovative, high-quality products and services.
- **AbbVie**: To discover and deliver advanced therapies that address some of the world's most complex and serious diseases.

### Four Areas of Focus

Transitioning from transactional to transformational marketing in the pharmaceutical industry, which is a value-based approach, involves, first and foremost, shifting the focus from short-term to long-term commercial and public health goals. The four key areas of focus are:

1. *Focus on Patient Education*: Pharma companies should educate patients and healthcare providers about different health conditions and treatment options, not just specific products they want to sell.
2. *Community Engagement*: Pharma companies would do better to engage with the communities they operate in to understand their needs and concerns, and to build trust by showing that they are committed to improving the health and well-being of patients.
3. *Support Independent Research*: Pharmaceutical companies should support independent research on their products to ensure that the information about safety and efficacy is reliable and unbiased.
4. *Follow Ethical Practices*: Pharma companies should follow ethical practices to ensure fair competition and support regulation to protect public health and safety.

## Making A Transition

How does a company make a transition from transactional to transformational marketing? By changing the focus from short-term sales goals to long-term public health goals and the current unethical practices to a sharper focus on becoming patient-centric by understanding their concerns and pain points and addressing them by providing relevant and meaningful patient education and engagement programs. The transition also involves a shift from unethical practices, such paying physicians cash, kickbacks, expensive gifts, and luxury holidays for prescribing their products to building enduring relationships based on mutual respect based on competence, and capabilities through scientific promotion and helping the HCPs provide better treatments to their patients.

## Vision to Transform

What it takes to make a transition from transactional to transformational marketing in the pharmaceutical industry, and what steps should it take? First and foremost, a company that wants to practice transformational marketing should have a vision of creating a deep and meaningful connection with its customers by delivering products and services beyond fulfilling basic needs and wants. The following ten steps would significantly help a pharmaceutical company's transformational journey in its marketing practices, regardless of its size— small, medium, or large enterprise.

1. *Create a clear and compelling vision*: The company must develop a vision that inspires and motivates employees and customers to strive for something greater than merely providing products and services that treat certain medical conditions. This vision must align with the company's values and mission and should be communicated effectively to all stakeholders so that they own it.
2. *Develop a deeper understanding of customers' needs and pain points*: The company must deeply understand its customers' emotional and psychological needs and should create products and services that address those needs. For example, it could involve deep listening using social technologies, and

conducting qualitative and quantitative consumer research, including ethnography, to gain insights into how customers think and consumers' pain points across their journeys.

3. *Be Authentic and Transparent*: The company must be authentic and transparent in its communication with all stakeholders. Remember, action speaks louder than words. Therefore, the company must walk the talk. It must be honest and open about its products, services, and values. Building trust with customers and stakeholders is essential to transformational marketing; authenticity and transparency help you do that.
4. *Innovate and Improve Continuously*: The customers' needs are continuously evolving, and therefore the company must innovate and improve its products and services to continue to meet the changing needs of its customers. This requires fostering a culture of experimentation and risk-taking, staying updated with the latest technologies, and conducting customer research.
5. *Develop Strong Leadership*: Transformational marketing requires strong leadership at all organizational levels, starting from the top. The leadership should be visionary and inspiring and should be supported by a culture that values innovation, customer focus, and continuous improvement.
6. *Be Flexible and Adaptable*: The marketplace is constantly and continuously changing. So, the company should be flexible and adaptable to changes in the market and customers' needs and should be able to pivot the strategy quickly as needed. Therefore, the company should adopt agile practices.
7. *Build Strong Partnerships*: In today's hyper-competitive marketplace, building partnerships with other companies, organizations, and customers can help the company to reach new audiences, gain access to new technologies and resources, and achieve its transformational marketing goals.
8. *Be Socially Responsible*: A strong Corporate Social Responsibility (CSR) program can help a company positively impact society and the environment, which can help build brand loyalty and trust with customers. It can also help to attract and retain new talent.

9. *Upgrade Technology*: Today, technology is not just a part of a business. It has become an integrated part of every small, medium, or large business. As a result, every business has become a technology business. Therefore, one must constantly and continuously upgrade technologically to stay ahead.
10. *Implement A Robust Digital Transformation Program*: Today, everything is already digital or becoming digital. So, without a doubt, digital is the way to go. However, it is important to put process before technology and customer before process while implementing a digital transformation.

## Technology Plays A Vital Role

In addition to focus and a change in mindset, technology plays a vital role in transitioning from transactional to transformational marketing in the pharmaceutical industry. Consider the following five areas where technology can help implement transformational marketing:

1. *Patient Engagement*: Technology can help create digital platforms, such as mobile apps, social media, and websites. These platforms can inform patients about their health conditions, treatment options, and side effects, gather feedback, and track patient outcomes.
2. *Big Data and Analytics*: Today, pharma companies can use technology to collect and analyze huge amounts of data from patients, healthcare providers, and other sources, which in turn helps companies better understand patient needs, pain points, and preferences to improve patient outcomes.
3. *Remote Patient Monitoring*: Pharma companies can use technology to remotely monitor patients' health status through wearables, telemedicine, and remote patient monitoring systems, enabling companies to understand patient outcomes better and to identify potential issues early on and intervene before they become serious problems.
4. *Clinical Trials*: Technology helps drug companies automate and streamline the clinical trial process through EDC (Electronic

Data Capture) Systems, ePRO (Electronic Patient Reported Outcomes), and other digital tools. Thus, technology can help reduce the time and costs associated with clinical trials and improve the data's accuracy and reliability.

5. *Artificial Intelligence and Machine Learning*: AI (Artificial Intelligence) and ML (Machine Learning) can help identify patterns and trends in patient data, predict patient outcomes, and identify at-risk patients. Thus, with AI and MI, pharma companies can tailor their marketing and engagement efforts to specific patient and HCP groups. Also, AI and ML can help pharma marketers identify patients who may be more likely to respond to specific treatments.

## What it Takes to Transform

Pharma's journey to transformational marketing involves pharmaceutical companies doing two things. One is to stop unethical and corrupt practices and start embracing ethical values and principles in running its business— in other words, pursue an ethical approach. It's not as difficult as it sounds. All it takes is to change the mindset, and that needs conviction. Next, one has to remove the 'Un' part of the 'Un-ethical' approach. That is stopping the corrupt practices like paying cash, kickbacks, expensive gifts, luxury holidays, and other instant gratification measures as a *quid pro quo* for prescriptions to physicians.

How does one embrace the ethical approach? By following scientific marketing principles to build trust, develop loyalty, and improve reputation. Start acquiring and nurturing new and relevant talent for the changing times and reinforce those by rewarding the new behaviors. By complying with the ethical codes and regulations governing the industry. By living up to ethical values and principles. By putting the patients at the center of everything you do. By being transparent in your communication, providing a balanced view of your products and services benefits and side effects, and publishing all your clinical trials. Above all, make sincere efforts to improve access to patients for your medicines because they are meant for them and would not serve the purpose if patients cannot access them.

The other is to embark on a company-wide digital transformation.

## Why Digital Transformation?

Digital transformation is becoming increasingly important for the pharmaceutical industry, allowing companies to streamline operations, improve efficiency and enhance customer experience. More specifically, there are three main reasons why pharma should embark on a company-wide digital transformation.

1. By implementing digital tools and technologies, companies can more easily track and monitor inventory, forecast demand, and optimize logistics. This can help reduce costs, improve delivery times, and ensure that the right products are available to customers when needed.
2. Digital transformation can help improve drug discovery and development. By leveraging advanced analytics and machine learning, pharmaceutical companies can more easily analyze large amounts of data, identify patterns, and make more informed decisions about which drugs to develop and how to bring them to market. This can lead to faster drug development times and more effective treatments.
3. In addition, digital transformation can also help pharmaceutical companies to improve their customer engagement and experience. By leveraging digital channels like social media, mobile apps, and websites, companies can more easily connect with patients, understand their needs, and provide tailored information and support. This can lead to improved patient outcomes, increased patient satisfaction, and greater brand loyalty.

Therefore, digital transformation is essential for the pharmaceutical industry to stay competitive, improve operations and customer engagement, and bring new, effective treatments to market more efficiently. In the era of technological advancements and digitalization, companies that do not embrace digital transformation risk falling behind their competitors.

The following section briefly outlines various technologies and frameworks essential in implementing a company-wide digital transformation strategy. These are:

1. Marketing Automation
2. Customer Relationship Management
3. Tablet Detailing
4. Virtual Engagement
5. Webinars
6. Website Engagement
7. Email Engagement
8. Mobile Engagement
9. Gamification
10. Multi and Omni Channel Engagement
11. Social Media Engagement
12. Search Engine Marketing
13. Artificial Intelligence (AI)
14. Machine Learning (ML)
15. Agile Marketing
16. Design Thinking
17. Content Marketing
18. Closed Loop Marketing
19. Data-Driven Marketing
20. One-on-One marketing (Personalization)
21. Cloud Computing
22. Augmented Reality
23. Virtual Reality
24. Blockchain in Pharma Marketing
25. Cybersecurity
26. Patient Engagement

# Marketing Automation

### What is Marketing Automation?

Marketing automation is a software that streamlines, automates, and measures marketing tasks and workflows to increase operational efficiency and help grow revenue faster. The essential components of a marketing automation system are :

A. *Central Marketing Database* is a place where you store all your marketing data, such as detailed prospect and customer interactions and behaviors, enabling you to segment and target the right message to each customer.

B. *Engagement Marketing Engine* is like an orchestra conductor and allows engagement marketing practices and conversations across online and offline channels through automation.

C. *Analytics* help you test, measure, and optimize marketing ROI (Return on Investment) and impact on revenue, helping you understand what worked, what didn't, and where you can improve.

D. MarTech is a collection of all effective, collaborative, and scalable marketing technology applications you need to help you achieve your goals. It's the place that helps you stay connected to your customers and stakeholders and align with them.

### What Can Marketing Automation Do?

Marketing automation can streamline, automate, and measure marketing tasks and workflow to increase operational efficiency and grow revenue faster. Thus marketing automation helps you optimize your business revenue and growth by aligning people, processes, and technology. In addition, it enables many modern marketing practices, such as:

- Lead generation and nurturing
- Segmentation
- Relationship marketing

- Cross-sell and Up-sell
- Customer Retention
- Measuring ROI
- Account-based marketing

John McTigue, a marketing expert, has this to say about marketing automation: *Without using automated marketing, you are just guessing and hoping that people will take the bait and be ready to buy your products. However, statistics show that buyers don't do that. Instead, they want to learn at their pace and be reached when they need more information or are ready to buy. A well-constructed marketing automation strategy makes that a reality.*

According to Forrester's Marketing Automation Technology Forecast, 2017 to 2023, marketing automation is one of the fastest-growing technologies. Initially used for large enterprises, marketing automation tools have become more prevalent and scalable for small and mid-sized businesses. Marketing automation is no longer a nice to have. It is essential. Consider these pointers, for example:

- Marketing automation technology is expected to show a 14 percent compounded annual growth rate (CAGR) over the next five years.
- The highest growth for 'through-channel marketing automation' platforms will reach 25% annually, with 'lead-to-revenue automation' platforms at 19.4 percent.

In addition, marketing automation helps marketers solve many problems. Consider these, for example:

- Marketers must do more with less in today's fast-paced, highly competitive business environment. Marketing automation frees up marketers' busy schedules with automated tasks and helps them focus on strategic work.
- Allows marketers to create, manage and automate marketing processes and conversations online and offline, optimizing marketing programs.

- Helps nurture prospects with relevant and useful information across multiple channels and can increase lead generation by 50 percent and lower cost per lead by 33 percent.
- Marketing automation can help marketers personalize customer journeys across all touchpoints.

**CRM and Marketing Automation: What's the Difference?**

While CRM (Customer Relationship Management) and Marketing Automation overlap, some key differences exist. Here are the major differences:

- Marketing automation targets people early in the sales and marketing funnel. For example, Your lead may have visited a website, clicked a link, or filled out a form, all of which help you gather data at an early stage of targeted marketing and automate many of the marketing tasks themselves.
- On the other hand, CRM is generally focused on sales and manages interactions with customers to facilitate conversions. It also handles existing customers and tracks their engagement.

CRM and Marketing automation are powerful tools that help marketers to gain and nurture leads, facilitate sales, and build and maintain brand loyalty, when used together.

## Customer Relationship Management (CRM)

The general perception of customer relationship management (CRM) in the Indian pharmaceutical industry is very narrow and transactional. It is based on customers' gratification through high-value gifts, foreign junkets, and even cash for physician prescriptions. This practice of transactional marketing is not CRM. What is CRM, then?

CRM is, first and foremost, a strategy. CRM is about identifying, satisfying, and maximizing the value of a company's best customers. CRM is about how companies interact and engage with customers at all the touch points throughout their journey. CRM is also about a set of software tools for customer relationship management. They should be implemented only after a well-defined strategy and operational plan with all the monitoring mechanisms. It has four major components:

1. *Operational CRM*: Stores customers' contact histories with all the relevant details so that sales and marketing people can retrieve them whenever needed. Operational CRM, as the name indicates, provides support to the sales and marketing teams (front office). This also includes customer data for managing campaigns and facilitates sales force automation.
2. *Analytical CRM*: Helps in analyzing customer data for designing and executing targeted marketing campaigns that are very useful and even essential in acquiring customers, cross-selling, up-selling, and making product-related decisions such as product augmentation and pricing.
3. *Sales Intelligence CRM*: Provides much-needed insights into customer acquisition and retention strategies in terms of planning, implementing, and monitoring customer-specific strategies.
4. *Collaborative CRM*: Customer service management is not just the marketing department's function alone. It requires teamwork. Several people from various other departments come into the picture in the customer service management process. Marketing is the front end of it. Collaborative CRM,

as the name suggests, covers all aspects of a company's dealings with customers that are handled by various departments within a company and shares it with the respective people as needed.

CRM, thus, is a process sponsored by the highest levels of an organization, embedded into the corporate culture, and pervades throughout the organization.

### 7 Benefits of Pharma CRM

The pharmaceutical industry is highly regulated and faces some unique challenges, such as how it can promote its products and services, comply with numerous regulations, generic competition, and marketing audit controls.

Customer Relationship Management system helps pharmaceutical companies with the ability to automate their processes, pinpoint marketing efforts, enhancing the effectiveness of their marketing teams. Danine Midura highlights seven benefits that Pharma CRM can offer for sales and marketing in her article, *7 Benefits of Pharma CRM for Sales & Marketing* published in the Technology Advisors blog on August 11, 2020. Here is a summary:

1. **Automated Processes** in a Pharma CRM help reps schedule physician visits, simplify other sales automation tools, and simplify the planning and execution of customer interactions, maximizing the marketing team's effectiveness. In addition, a unified database of products with detailed descriptions of each drug and their clinical trial details acts as a powerful reference point for reps on the go, enabling them to develop product segment matrices and plan better sales calls. Furthermore, CRM's automatic data imports deliver useful perspectives on how sales teams manage their work with customers and other stakeholders help analyze the results of marketing efforts to improve or change to optimize the results.
2. **Integration with Industry Databases:** Integrating CRM with the company's IT infrastructures, such as ERP (Enterprise Resource Planning), HR Solutions, and support centers helps

remove the data silos, improve collaboration between teams, and amplify the benefits that data analytics can offer.

3. **Customer Journey Tracking:** CRM software helps marketers track the customer journey. Customer journey tracking informs you where the customer stands according to a set of interactions with your brand, documenting the full experience of that customer. Based on the customer's interactions and communication history, sales reps with access to the CRM can quickly understand where that customer is within their customer journey lifecycle and adapt their communications to be more relevant and proactive to their needs.
4. **Mobile CRM:** Mobile CRM accessibility has become the standard in almost all pharma CRM solutions as pharmaceutical sales reps need to access information on the go. Mobile CRM enables these representatives to quickly reference customer histories, keep notes on important discussions, review and log phone calls and emails, and look at recent customer touchpoints before and after a meeting. These insights help pharma reps build better customer relationships and have more productive sales interactions.
5. **Data Augmentation**: Data augmentation or enhancement helps pharma sales reps pull data about prospects from all over the Web with minimal primary information, such as the prospect's name and email address. CRM can fill in the blanks on the prospect's social media accounts, job titles, and company mentions in the news and help create robust profiles to inform and help foster more personalized interactions with customers.
6. **Targeted Marketing:** Pharma CRM helps marketers gauge the current market conditions based on customer behavior and buying habits and helps them realign markets by territory, product, segmentation categories, or even response to new regulations. In addition, CRM's functions, like automated campaign management, keep marketing rolling forward with integrated data sets, helping marketers predict emerging trends so marketing can stay one step ahead of the competition.

7. **Activity Tracking:** Pharma CRM solutions can capture leads from every touch point and aggregate that information in a single location. The data aggregated by CRM enables pharma companies to capture more information and accurately track leads and understand the most effective channels for engaging with them.

Thus, Pharma CRM is a powerful tool that reveals deeper data patterns and streamlines the sales process by aligning other departments.

## Tablet Detailing

For a long time, pharmaceutical companies have been using paper-based detail aids to promote their products to physicians. Later, pharma companies started using laptops for detailing their products to physicians to gain better attention. While the laptop detailing has some advantages over the paper detail-aids or visual aids regarding audio and video content and some graphic links, it was not very convenient to use. The time taken for the laptop to boot was longer, and the navigation was challenging. Furthermore, internet speed also was grossly inadequate.

On the other hand, the laptops were suitable for offline presentations of PowerPoint and videos. The detail aids, as a result, were mere electronic versions of paper-based detail aids or visual aids. Some multinational and large Indian pharma companies used laptops to detail their new and significant products with their managers in the field.

However, the introduction of the iPad by Apple in 2010 changed all that. iPad, by its capabilities, has become the most convenient, versatile, and useful tool for detailing, and the pharmaceutical industry quickly embraced it. The iPad has achieved a phenomenal adoption rate within six years of its introduction. Over 80 percent of Pharma companies use it as their device for detailing their products in North America and some European countries. While iPad has become the numero uno digital tactic for the pharma industry, some companies have started using other tablets too as detail aids. The share of other tablets, however, is meager.

### Why is iPad Pharma's Device of Choice?

Pharma has embraced the iPad as their device of choice for detailing by their sales force as it offers many advantages. Consider these for example:

A. Excellent visual quality with a large screen for delivering video, graphics, and animation content
B. The sleek touchscreen and lightweight make it easy to carry

C. 4G, or 5G wireless connectivity is built into the device for speedy internet connection.

D. Capability to combine a Customer Relationship Management (CRM) system that includes content visualizing, call planning, signature capture, tracking, scheduling, showing videos, presenting detail-aids, GPS, integrated e-mail, and Closed-Loop-Marketing (CLM) system into the device, making it an ideal tool for sales force automation (SFA)

Furthermore, on iPad, it is also possible to do sales management side of sales force effectiveness, such as territory planning, sales operation, sales forecasting, promotion response planning analysis, and some return on investment (ROI) calculations and marketing planning. Besides, the Pharma industry can measure how sales campaigns get the most response rates or lead with the data input captured through the closed-loop-marketing system on iPad.

### 3 Ways to Get the Best Out of Your iPad Detailing

Here are three essential success factors that can significantly improve your sales reps' performance of iPad detailing.

1. Use the iPad detail aid to support the verbal conversations of your sales reps. Spelling out everything leads to an overload of presentation slides or PDF pages. Experienced Pharma sales reps used the printed detail-aids to strengthen their detail talk and never read the text verbatim. Likewise, iPad detail-aid, too, should make the conversation impactful and not replace it. A good tablet detail-aid helps strengthen the delivery of the message with a logical structure, examples, and cases and support to the product story. The rep should always lead the conversation.
2. Stop broadcasting your message and start engaging. The tablet is ideal for capturing data on treatment practices and brand perception of physicians. Continuous content refinement will get the attention and, more importantly, sustain the attention of HCPs (Healthcare Professionals) and reinforce Pharma sales representatives' motivation.

3. Maximize the graphical opportunities of the iPad to strengthen the marketing message. iPad and other tablets have enormous potential to go beyond text and images. You can, for example, include treatment class, product mode of action animations, video statements by key opinion leaders (KOLs), patient cases, and much more. You can make the content come alive and make the detail-talk interactive, and memorable experience with the help of the iPad.

### Common Mistakes To Avoid

The three most common mistakes to avoid while implementing tablet detailing are:

1. Do not create a digital version of a paper visual aid. The early adopters of iPad detailing made the mistake of creating exact copies of their visual aids without utilizing any additional functionalities such as video, animation, and data capturing that iPad has.
2. Do not overload the iPad with too much content. It is possible to add almost an infinite stream of content. Therefore, it is tempting to equip the sales force with almost encyclopedic information to handle any objection and question. That will only make it difficult, if not impossible, for the sales reps to search for the relevant content during the 'detail call.' Furthermore, it is observed that most reps use only three to five pages or slides of tablet detail aid to support their conversation. However, when you want to add more content, make sure it is not a part of the main navigation stream and add it to a sub-menu. Always ensure that the main detail-talk flows are short and to the point. Let the rest of the content be a part of the sub-menu.
3. Do not overdo it. Never use animation for the sake of animation video content just because it can be done well on an iPad. Use them only when they are strengthening the key messages.

## Implementing iPad Detailing Effectively

Many sales reps feel somewhat uncomfortable about the tablet detailing platform because one can monitor— if and how much the detail aid is being used in the field. Therefore, you need to explain the benefits of iPad detailing in making their messages more effective and their engagement with the physicians much better.

Remember to involve your sales reps immediately when launching a tablet detailing program. As they have to use the device and the detail aids daily, they are best positioned to provide valuable input on what is required to make tablet detailing a success.

Leverage the call expertise of your sales reps and show them how iPad detailing can support their conversations with the physicians and take it to the next level. The best way is to demonstrate how it will help them engage physicians, which is getting increasingly difficult.

Train your sales reps on how and when to use the iPad or any other tablet that you may be using. Use role-play as part of the training exercise, record the role-play session, and show it to reps to precisely know where and what they need to correct. It is observed that many representatives today look more at the visual aids rather than at the physician they are detailing to. The same is the case with the sales reps who were early adopters of tablet detailing. It is observed that they, too, were looking more at the iPad than at the doctor they were presenting to.

## Winning with iPad Detailing

The iPad has become the number one digital tool for over eighty percent of pharmaceutical companies in North America and some European countries. However, many have yet to unleash the iPad's potential to engage physicians and healthcare professionals. In addition, the iPad has the potential to be the single device for sales force automation and customer relationship management, and closed-loop marketing for the pharmaceutical sales force.

Closed-loop marketing captures information on every interaction with the physicians regardless of the channel being used. It acts on the information gathered in designing subsequent messages, thus ensuring target-specific and highly personalized marketing messages. Tablet detailing can be an excellent tool to start the closed-loop marketing experience with additional channels integrated later.

### Designing and Delivering Segment-Specific Content

Your iPad can deliver the content specific to in each physician segment across the stages of their adoption of your product. Stages of adoption, also called the *ladder of adoption*, are the step-wise overview of a doctor's possible steps when considering a brand. Consider the following segments (based on stages in which the physicians currently are) for example:

1. Physicians who are unaware of the treatment or brand or using a different brand
2. Physicians who have just started prescribing and sent their first prescriptions
3. Physicians who have taken up your brand and prescribing more or less regularly
4. Physicians, who are high-volume prescribers and brand advocates who influence peers proactively to prescribe the brand

Each segment requires a different marketing message to meet its communication needs. For example, during their calls, the sales representatives will try to determine the stage at which their physicians are. Brand managers design specific communication messages to meet all physicians' communication needs as per their relative positions in the stages of adoption. The sales representatives can thus quickly access the right information and the required presentation slide decks or videos on their tablets, and deliver engaging detail. Without the segment-specific content, it would be much harder for the sales force to use their tablets effectively and engagingly.

### How to Create an Engaging 'Tablet Detail'

Brand managers usually create tablet-detailing communication strategies used by their sales forces. Here are four major areas to consider when designing engaging tablet details:

1. *Target Audience Segmentation*: You need to cover all customer segments in your detail aid, including your customers' position on the adoption ladder. Also, pre-structure the detailing talk for each position on the adoption ladder. When the sales rep identifies the position of a physician on the ladder of adoption during a call, he can select the appropriate detail that best fits the customer's position on the adoption ladder. Without segmented content, sales reps will find it difficult to use the tablet effectively, thus limiting the impact of the call.
2. *Creating Segment-Specific Content*: Once you define the segments and relative positions your targeted physicians may occupy on the adoption ladder, tailor all content according to the behavioral objectives (what you want the physicians to do after the call) and messages. It increases the interest of the physicians and motivates your sales reps to use the tablet detail. There may be many messages that apply to these segments. However, it is important to limit the messages between two and three. More may be confusing for the sales force, and make a physician call more complex.
3. *Different Content Types to Suit Different Customer Profiles*: While it is possible to create many different types of content in terms of graphics, videos, animations, and methods, all the content may not be liked by all physicians. Some may prefer medical, scientific content, such as guidelines and publication downloads, while others may prefer more patient-oriented content, such as patient cases and support programs. Differentiating customers according to their interests is the key, so your sales force can quickly choose a content piece that matches the physician's interests and preferences.
4. *Enabling for a Successful Closure*: Closing the call is perhaps the most difficult part of the conversation. Therefore, a strong

closure screen will be a useful tool. It could consist of presenting different patient profiles, or suggesting to the physician to agree to prescribe a product to a new patient profile before the next call. Other possibilities are that a sales rep could summarize the conversation or invite the physician to a follow-up activity. A strong closure sets up a future interaction, which could also be delivered by another channel than the sales representative.

## Improving Tablet Presentation Skills

While the iPad is a versatile digital tool for effectively engaging physicians during a detail, it requires considerable skills and practice. Training to enhance presentation skills by using a tablet is essential. Consider training your sales force in these areas:

1. When to use an iPad, and when not to use
2. How to segment physicians
3. How to use and handle the iPad properly
4. How to properly use the iPad? A representative, for example, should not be looking at his tablet while presenting a detailed talk but should be looking at the physician.
5. Which message to address first, preferably for each segment?
6. How to close the conversation?
7. What to do if no segmentation can be determined?
8. What to do if the physician gives only 30 seconds? What is his elevator pitch?

Elementary and straightforward as they may seem, these basics are essential as they make the representative resourceful regarding conducting themselves in any situation.

The critical question is, how are you using your iPad or tablet? The effectiveness of iPad or tablet detailing depends on whether you are trying to unleash its full potential. David Ormesher, CEO of Closerlook, one of the leading digital marketing consultancy firms in the US, puts it succinctly when he says: *If the tablet is not part of an authentic, closed-loop strategy, it simply replaces a glossy presentation with a shiny one. It is like replacing a traditional hammer with an auto-hammer: it's cool, and it's electric, but in the end, it still does the same thing.*

**Tablet Detailing: Future Outlook!**

The Indian Pharma companies are rather slow to adopt tablet detailing. The continued heavy reliance on the transactional business model and quick success tactics based on massive rewards—monetary, expensive gifts, and substantial discounts for purchasing physicians may be one of the main reasons. The few companies which are investing in technology are the ones that are implementing tablet detailing. However, tablet detailing in India is currently a digital version of the paper-based visual aid with a few bells and whistles. It is not yet exploiting the full capabilities that tablets, in general, and iPad, in particular, can offer. It is like a glossy presentation being replaced by a shiny one.

As competition intensifies and when the current arms race of discount and gratification-based selling begins to recede because they no longer can deliver, Pharma companies that have ignored the telltale signs of the changing landscape of the marketplace would realize the need to invest in technology, development, and the implement the best practices. The innovators and early adopters of technology are sure to reap the rich rewards in the changing marketplace.

# Virtual Engagement

Doctors often rely on pharmaceutical companies for drug information. However, for the past few years, doctors could not find time to meet with Pharma sales reps for various reasons. In addition, many large hospitals and Group Practices have been restricting their access to pharmaceutical companies.

## COVID-19 Drives Virtual Engagement of Customers

The recent global pandemic, COVID-19, has caused many biopharmaceutical companies to rethink their customer engagement practices more critically. Although well underway before COVID-19, the pandemic accelerated the pace of change in how they operate and engage their customers. As a result, the pharmaceutical industry that relied on the traditional face-to-face sales force and physicians' interactions before the pandemic was forced to change significantly by reinforcing virtual communication approaches.

From the start of the pandemic traditional in-person sales reps' visits sharply declined due to social distancing and lockdown measures. As a result, pharma companies had to adapt rapidly to this new world, and start or increase the process of engaging their customers virtually. In addition, they had to train their sales reps to use virtual tools to stay connected with physicians. As the restrictions around in-person interactions and hospital visits continued, the relationships between physicians and pharma sales reps were changing, creating a substantial increase in demand for virtual and online meeting software tools.

According to a GlobalData poll conducted with pharmaceutical industry professionals in April and May 2020, three-fourths of the 456 respondents opined virtual interactions remain either a stand-alone option or a mix of in-person and virtual interactions even after COVID-19 recedes.

When asked if he prefers remote or online visits at an Online Veeva Summit, Dr. Andrew J. Moore, hematologist, and oncologist at Southeast Cancer Center, said *digital is the way to go. The biggest impact COVID has had on our clinics is the efficiency and workflow of our day. Therefore, I would like to see virtual visits continue absolutely.*

### Video Calls Preferred

In another GlobalData poll completed with healthcare providers in the same period as physicians this time, about thirty percent of the 317 respondents believed that video calls are the best tools for virtual interactions between sales reps and HCPs, followed by messaging (20 percent) and phone calls (16 percent). Many respondents saw video calls as the best alternative to traditional in-person interactions with pharma sales reps. Although video calls cannot fully capture body language, they can still provide information and facial expressions, giving some non-verbal cues to help establish rapport.

### Virtual Engagement is the Way!

Every way you look, virtual engagement seems to create a sustainable engagement with stakeholders in the pharmaceutical industry.

Virtual or remote engagement is sometimes referred to as *tele-detailing*, a *web call*, *eDetailing*, *eRep*, or *eMedical Science Liaison*, depending on whether the presenter is a sales rep or a medical science liaison (MSL) interacting with the customer. Virtual engagement is an interactive, real-time meeting with a healthcare provider online.

### Virtual Engagement Adoption Increasing

Changing customer behavior drives the increasing adoption rate of virtual engagement in the pharmaceutical industry. Here are the main reasons:

1. Access to physicians is becoming increasingly difficult and has come to a grinding halt with COVID-19, resulting in a drastic decline in sales rep productivity. These trends are increasing pressure on the pharma industry, and virtual engagement could be a good alternative to extend reach and contact frequency to sustain the business.

2. Virtual engagement allows physicians to interact from the comfort of their homes, and as such, physicians are also more prepared to spend time with sales reps and MSLs (Medical Science Liaisons).
3. Since 2014, when the tipping point between digital immigrants and digital natives among physicians was reached, more than sixty percent of doctors today are very familiar with digital communication channels, such as Skype, FaceTime, and Zoom, in their personal lives.

When considering all these factors, virtual sales reps and physicians' engagement will likely remain a new standard even after COVID-19 recedes.

### Factors Influencing Virtual Engagement

There seems to be a strong relationship between remote participation and specific parameters.

1. Age. Healthcare professionals participating in virtual engagement are slightly younger than physicians on average.
2. Physicians who own a tablet or smartphone are likelier to accept an invitation for remote engagement.
3. A relationship with a sales rep or MSL (Medical Science Liaison) makes a difference. Physicians are three times more likely to accept an invitation to interact remotely if they already have a good relationship with the sales rep or medical science liaison.
4. Physicians who visit pharma-owned websites or are active on social networks for professional purposes are more likely to accept an invitation than those who don't.

### Tele-Detailing

Tele-detailing allows drug companies to provide this interaction remotely on the doctors' terms. Tele-detailing (also known as 'e-Detailing), or 'remote or virtual detailing' or 'web calls,' is engaging remotely with doctors and Healthcare Professionals (HCPs). It is an interactive, online, real-time meeting with a physician. As physician access for Pharma companies is increasingly getting restricted, Tele-detailing or e-detailing help extend reach and contact frequency to create a more significant impact.

### Two Basic Types

There are two basic types of eDetailing. One is Interactive (Virtual or Remote) detailing. A physician can access the interactive detailing at his convenience. A typical virtual eDetailing call contains multimedia content such as a slide deck, video, or animation that lasts between 5 and 15 minutes. It is mainly used by many pharmaceutical companies in North America and some European countries. However, its adoption is increasing in many emerging markets, such as India, China, and Brazil. Virtual detailing offers an advantage to the physician as he can control the detailing time and the contents he would like to access. He can also request face-to-face detailing with a company sales representative, and samples of the products he would want to try at the end of the detailing session.

Live or video detailing, on the contrary, is a face-to-face video conference with the physician speaking to a pharmaceutical sales representative on a smartphone, a tablet such as an iPad, or a computer webcam. The Pharma sales rep directs the detailing and leads him through the presentation containing similar multimedia elements as virtual or live detailing. It is more or less similar to traditional detailing, except that it is a video and not real.

### Three Models

Currently, the pharmaceutical industry follows three models for implementing e-Detailing.

1. The *Outsourced Model*, where an external agency manages all components and activities of e-Detailing of a company. External sales reps are contracted to make the remote detail calls.
2. The *Dual Model* is where an external agency conducts most activities, such as scheduling and arranging appointments with physicians, except the actual call delivery.
3. The *Hybrid Model* is where an external agency does the initial setup, training, and platform configuration. The call scheduling

and call delivery are done internally by the company employees.

### Implementing e-Detailing

The three necessary steps in implementing an e-detailing program are scheduling a meeting, running the meeting (e-Detail), and evaluating the meeting.

1. Scheduling a meeting involves taking a prior appointment with the physician when he would like to log in at the appointed website to participate in the company's e-Detail.
2. Running a meeting presents the e-Detail by the Pharma company's representative or MSL if the meeting is about only clinical or medical issues, not the product. The meeting takes place over Skype, FaceTime, or a similar web-based platform or at the designated website.
3. Evaluating e-Detail is the final step in the process. It is for collecting essential feedback to improve the future e-Details and provide the physician with additional information such as clinical trial reports, web links, and samples.

Following the e-Detail, most physicians would like new product information, educational programs, links to other sources, and drug formulary updates (mainly for the reimbursement status of the prescribed products).

### e-Detailing Strategies

e-Detailing offers many advantages to a pharmaceutical company. The most common strategies followed by Pharma companies are:

1. *Coverage* means extending the reach toward customer groups currently not seen by the field force.
2. *Frequency*, the number of calls a Pharma sales rep makes, can be increased compared to face-to-face calls as there are no commuting and waiting times in web-based, remote calls.
3. *Vacant Territories* can be covered adequately.

4. Companies can offer *Services On-Demand* to physicians by addressing their online requests for product or clinical information and samples.
5. E-Detailing makes *Differentiated Messaging* possible. With e-Detailing, you can target your message specific to the individual physicians based on the feedback from the previous interactions, whether face-to-face or e-Details.

**Objectives of e-Detailing**

It is essential to set clear goals before implementing an e-detailing program. Consider these objectives, for example:

1. To increase the length or duration of an e-Detail. When the e-detail is more interactive and intuitive, it is more enjoyable with appropriate animation and video built into the communication.
2. To increase the effectiveness of e-Detail by delivering personalized, target-specific communication relevant to each physician.
3. To decrease the cost of the sales force without losing effectiveness.

**Training is Important**

Prior training of the Pharma sales representatives who deliver e-Detailing is vital as it involves new technology. The training focuses mainly on two broad areas. One is on how to present the content using the technology right, and how to manipulate the webcam and the audio headset, give the physician access to the meeting, and share the relevant content securely. The second aspect of the training focus is on remote communication. Pharma reps will often not be able to see the physician during an e-detail call and still have to ensure that physicians pay attention to them instead of doing something else during the call.

**Three Key Factors for detailing Success**

Three main factors influence the success of e-Detailing. One is the cost. On average, per physician, the e-Detailing cost is estimated to be between US $100 - 150 compared to the face-to-face detailing cost of US $200 - 250 per physician. Secondly, e-Detailing is more effective. The duration of an e-Detail is about 10 minutes on average compared to 2 minutes for a face-to-face detail. Increasing physician acceptance is the third success factor. A quintiles study observed that a whopping 97 percent of physicians are likely to repeat an e-Detail. Almost three-quarters of them agree that e-Detailing is the future of pharmaceutical contact.

### Why Do Physicians Like e-Detailing?

Here are some reasons why physicians like e-detailing.

1. Ease of scheduling.
2. Fits into the physicians' schedules better
3. Convenient, as they have the option to take the detail at the office or home.
4. Saves time.

### e-Detailing in Branded-Generics Markets

So far, we have discussed e-Detailing in research-based Pharma markets such as the US and Europe. However, as discussed here, E-detailing is not yet practiced by the Indian pharmaceutical industry. One primary reason could be that e-Detailing is not a one-size-fits-all strategy. E-Detailing is more beneficial in a product lifecycle's pre-launch and post-launch phases. In the pre-launch phase, to create and increase awareness, and in the post-launch to accelerate the adoption rate. During these stages, there is a great need for product and clinical information. Perhaps, marketers in branded-generics markets may not yet feel a pressing need for e-Detailing. The current transactional model of pharmaceutical selling, which is based more on gratification, and the consequent complacent attitude by some Pharma marketers, maybe another reason.

That said, e-Detailing is undoubtedly important as it offers the same advantages in branded-generics markets as in research-based Pharma markets. The new product launches are crucial for companies in branded-generics markets too. Moreover, physician access is increasingly getting restricted, even in branded generic markets such as India. e-Detailing offers a cost-effective solution to reach the difficult-to-see, or even no-see physicians.

Pharma Marketers, who are progressive and forward-looking, would be better off understanding the best practices of e-Detailing to stay ahead of the curve and reap the rewards of e-Detailing.

**Virtual Engagement: Future Outlook**

Mark McLaughlin and Brian Mahoney outline the future outlook for virtual engagement in the Pharmaceutical Executive article, *Opening Digital Doors: Remote Selling During COVID-19*. Here is the summation of what they say:

1. Relationship selling will continue to be important. However, reps must also operate in a digital, virtual environment, given the dramatic shifts in face-to-face access and HCP preferences to digital.
2. Digital opens the door for reps to become more flexible and engage more effectively to deliver what the healthcare professionals need at any given moment.
3. As the industry moves ahead, digital channels will complement in-person visits in a significant way. From remote meetings to email, virtual engagement strategies will become more useful in improving and sustaining customer relations.
4. Biopharma companies have finally found the flexibility to engage customers on their terms and get doctors the information they need, when they need it, and how they need it. The industry is entering its greatest era of effectiveness and efficiency in serving its many stakeholders.

# Webinars

What is a Webinar? A webinar is an interactive presentation given to a large, geographically dispersed audience who can engage with the presenter and each other in some way. A webinar is more like an interactive seminar conducted on the internet. For example, a webinar may contain audience polls, Q&A (Question&Answer) chat functions, and whiteboards.

A webcast is a one-to-many presentation conducted on the internet. It is much like a traditional TV broadcast but shown live via an internet stream. The recording is then stored on a web page for future viewing.

Webinars are one of the most effective channels in the pharmaceutical industry today because they can extend reach and frequency. Besides, they can save customer time and money and offer a convenient means of receiving relevant communications. As a result, healthcare professionals (HCPs) can consume more educational material over time in bite-size chunks that are accessible with the ability to replay the content right from their office, house, or even on their smartphones.

### Use of Webinars in Pharma

Pharmaceutical companies predominantly use medical education webinars and engage HCPs in research or treatment discussions. They are also great for connecting HCPs with key opinion leaders (KOLs) more frequently. Pharmaceutical companies may also use webinars for brand promotion, as they are not the main drivers of the channel's success. Hosting a webinar requires using specialized software that connects the presenter and audience PCs (personal computers), tablets, or smartphones. Above all, a webinar requires the most suitable content. Therefore, good, relevant content remains the core of a webinar's success.

Webinar channels can provide a significant opportunity to showcase your organization's expertise in a disease area or a portfolio of products and create or renew interest among your targeted HCPs.

So what is the formula for winning through webinars? Start with understanding your target customers' journey and ensure the relevant content they value and deliver an engaging experience.

**How to Conduct Engaging Webinars?**

How to get the most out of your webinars? How to conduct highly engaging webinars? The key to providing more engaging and interactive webinars is to allow attendees to personalize their experiences. When busy physicians agree to exchange their precious time for informative content, they expect to find the value they could not readily receive otherwise. Targeting the right audience and posting excellent pre and post-marketing of the event, know-how, and using the best technology, determine the success rate of your seminar.

Content is the king, even in webinars. Your content should provide the value that they are seeking. Gurpinder Singh, Head of Digital Marketing at GlaxoSmithKline, India, offers valuable tips to improve engagement and involvement in his insightful article titled, *Webinars in pharmaceutical marketing*, published in the LinkedIn blog on June 18, 2018.

1. *Planning your webinar*: Planning is crucial. Before you plan your webinar, it is essential to create a detailed plan and define key performance indicators (KPIs) that you will measure in the webinar. Proper planning improves the chances of success significantly.
2. *Run surveys*: Ask your customers and find out their pain points and wants. Your webinar topics should address these pains and wants. Remember to ask your customers the right time and day for the webinar.
3. *Choose the right platform*: Choosing the right platform to run your webinar is very important. Your audience will stay long from various locations and multiple devices, where they might experience fluctuations in internet connectivity. Therefore, ensure the platform supports low-bandwidth streaming and is viewable on devices without additional plug-ins.

4. *Bring the expert speaker*: Once you decide on the topic, engage the expert who can deliver value to your audience. Ensure the speaker is aware of the surveys you ran so they know the pains and wants of the target audience to deliver a highly relevant and engaging presentation.
5. *Promote your webinar*: Your webinar is not an isolated event. It is a vital part of your marketing campaign. It is the core of your content strategy, and you must promote it appropriately across all your marketing channels. Start your promoting your webinar at least one month before the event. Create a landing page for registration and spread the word about the event using all possible channels. Make a detailed communication plan for your webinars. Here are some crucial points to ponder and act upon while promoting your webinar.
6. Record a one-minute video of the expert speaker to describe what to expect from the webinar.
7. Send the initial invite 3 to 4 weeks before the webinar. This communication should have similar content to the landing page, letting your target audience know about the webinar – when it is and giving them an effortless way to register. Let every webinar communication include the following text: *Can't attend the webinar? Register, and we'll send you the recording*.
8. Send a targeted reminder just two weeks before the event. This communication is for the audiences who opened your invite email but did not register. It should remind them about the upcoming webinar and ask for feedback on why they haven't registered yet. Embed a survey in the email asking for their reasons. For example— maybe the day or time doesn't work for them, or perhaps they're not interested in that topic. Whatever the reason is, you can take this feedback and use it to plan your next webinar. Use SMS (Short Messaging Service) as one of the channels in your webinar marketing campaign. Mobile has become the most convenient channel for communicating transnational information and marketing messages. Now, you need to send the reminder just one day before or on the day of the webinar. Once you have the desired

number of HCPs registered doesn't mean they will show up. If you promote your webinar 2 to 4 weeks in advance, most registered HCPs will likely forget the date on which the webinar goes live. Remember to send out reminder emails or SMS one day before and on the day of the live event to ensure a high attendance rate.

Here are five important points to consider while conducting an engaging webinar:

1. *What is the right duration of the webinar*: Webinar audiences prefer attending webinars that run between 30 to 45 minutes and not longer. Plan your webinar, therefore, no longer than 30 to 45 minutes. If you are dealing with a complex topic, break it down, run a series of webinars and ensure that the related information is available on your website.
2. *Keep it interactive*: Use Q & A (Question & Answer) and polls throughout the presentation to get a feel of your audience's engagement level and to keep them involved. See if you can offer to download or read more material during the event. Think of any VAS (Value Added Service) you can provide during the live event. Let your customers feel the value of attending live versus on-demand.
3. *Start and end on time*: Be respectful of your audience's time by starting and ending at the designated times.
4. *Ensure your webinar is viewable on-demand*: Please make the on-demand available as soon as possible for customers unable to attend the live version.
5. *Marketing after the webinar*: Do not let the webinars you have created end on the event date. Since you have created valuable content, plan a post-webinar communication proactively for HCPs who attended or missed the webinar. Plan, for example, a summary of the webinar and some FAQs (Frequently Asked Questions) based on the Q&A session, along with the on-demand link, which your customers will perceive as an excellent value. Also, share a calendar of all past and future webinars with your engaged customers.

## Pharma's Webinar Strategy

How can pharmaceutical companies benefit from webinars? Many pharma marketers recognized that they must go beyond banner and search ads and reinvent their strategy. Likewise, pharmaceutical companies must broaden their vision to gain insights from the HCP community. Here are three important reasons why a webinar channel is best suited to provide these insights and reach its target audience with highly relevant and valuable content.

1. Webinars offer the best real-time exchange of queries from physicians. Doctors tend to ask questions during webinars, which Pharma can analyze and effectively communicate to their target audience with the best solutions.
2. Webinars help pharmaceutical companies identify clinical dilemmas and address them immediately. Through webinars, Pharma can understand major clinical dilemmas from the specialties they want to promote their products and deliver the solutions they have already identified to a large audience through key opinion leaders (KOLs) convening webinars.
3. Webinars help pharmaceutical companies earn recognition as knowledge partners and bring a significant shift in the perception of the physician community from the current banner and flyer marketers. The relationship with HCPs as a knowledge partner is an excellent opportunity for Pharma to project its values and communicate its outstanding contribution to healthcare.

## Effective Engagement

How should Pharma engage HCPs in Webinars? Phanish Chandra, CEO of Docplexus, a leading doctors' network from India, offers three valuable suggestions for engaging with HCPs through webinars in his Docplexus blog article titled, *How to be successful with KOL webinar*, published on July 5, 2016. As he rightly says, Pharma already has a lot of clinical insights and resources to conduct webinars on a vast range of topics. However, it needs a comprehensive strategy to unleash the power of the webinar channel.

He suggests that Pharma's webinar strategy should incorporate the following points.

A. *Identify key therapeutic areas and novel treatments*: Pharma should identify key therapeutic areas where most practitioners face practical problems. It will help them unearth the unmet needs of doctors. Also, Pharma should identify any discovery or solution they can offer to address the problems of practicing physicians.

B. *Address the questions of the physician fraternity by recognized KOL (Key Opinion Leader)*: Identify key opinion leaders (KOLs) in target specialties to present pertinent topics in that specialty. Doctors are inclined to learn from recognized peers in their community. Furthermore, eminent KOLs can help resolve the dilemmas that practitioners face and address their questions immediately, thus enabling other doctors to better patient care.

C. *Provide additional resources to doctors*: In addition to live streaming of their webinars, Pharma should provide other resources like presentations or videos to help the audience better equip themselves with clinical knowledge. Pharmaceutical companies can also promote their proprietary insights, which can, in turn, change physicians' prescription behavior.

## Website Engagement

A website is the foundation of digital marketing. For many companies, the website (mobile optimized) is the heart of their digital strategy. It is the hub to which all other marketing activities are directed. According to Inbound Marketing Best Practices, 60 percent of all leads generated should come from inbound traffic, and that traffic needs a home. The website is that home.

In the early days of the internet, a website was more like a simple online business presentation. However, it has evolved from a one-way promotional channel to an integrated two-way service and educational platform. Today, a website can fulfill many functions in pharmaceutical companies. Consider these, for example:

1. **Corporate Websites** give general company information, information for investors, and information about employees. Often, It has limited information for Healthcare Professionals (HCPs) and patients. Some pharmaceutical companies have a comprehensive website with multiple sections containing information about their products, research pipeline, and other activities.
2. **Product Websites** inform HCPs about specific products giving all the relevant product information.
3. **Disease Websites** inform patients about a disease condition, the possible treatment options, and the implications of not starting treatment.
4. As the name suggests, **Campaign Websites** are separate websites for specific campaigns, although temporary. However, the campaign websites of late are becoming less popular. Pharmaceutical companies seem to be making their main websites flexible enough to cover the campaigns.
5. **The HCP Self-service Portals** are more comprehensive sites providing full service across products with features that help develop 360-degree personalization. For HCP websites, validation is required in most countries. Two options are commonly used for validation. One is to embed a third-party

validation tool in your website. The other is a validation system run by the pharmaceutical company itself. The advantage of the company's system is that the validation takes place against the CRM database, based on a few fields such as HCP's name and license number. Whatever option you choose, make sure you make the validation process hassle-free.

**How to Build An Engaging Website?**

To build an engaging website, one needs to define the strategy and goals of the proposed website and its target audience. The key questions to ask when building an engaging website are:

1. Who is the desired target audience for the website? For whom is this website being built?
2. How do we know that the website is engaging enough? Successful? What will success look like?
3. Which KPIs (Key Performance Indicators) should we monitor and use?
4. What are the conversion points? What is the desired behavior that we want from our visitors? What do we want the visitor to do when she visits our site?

Only after answering these questions should the site structure and content be defined. These processes are highly interrelated. For instance, a series of videos will require a different site structure than a library of text documents. In addition, we must keep our websites updated and fresh; therefore, we need the plan to update the website with new content continuously.

Creating the design of the website and building it are the next steps. These should happen in tandem with continuous user testing and testing for bugs. Now, the site is ready for launch. However, a website launch is not the endpoint of a website journey. Four follow-up actions are crucial:

- Firstly, you must ensure that the site is well-promoted.
- Secondly, measure all traffic coming into the website and monitor the usage pattern of visitors.

- Thirdly, build up further content based on usage results. Also, continuously optimize and maintain the site.
- Ensure feedback on all these points into a continuous learning process, which you can use to refine the strategy, KPIs (Key Performance Indicators), content, and editorial plans.

Continuous Updating of the website can be a challenging task. Therefore, create both pre and post-launch content plans before launching the site.

Also, to optimize your website engagement, ensure that you:

- Create multiple types of content and include personalization and segmentation options to keep your website robust after launch.
- Integrate your website into your overall content marketing plan rather than leaving it as a separate entity.

**Measure What Matters**

Websites can be great trackers. They have a significant advantage in that you can track almost everything you want to. There are three types of metrics that are important. They are Exposure metrics, Interaction metrics, and Engagement metrics. You can develop KPIs for each of these metrics.

1. *Exposure Metrics* looks at your website's reach by measuring the number of visitors you can draw or the average number of pages they visit.
2. *Interaction Metrics* access specific content items you have identified as key for your visitors to interact with.
3. *Engagement Metrics* look at how optimal the user experience is for your website visitors. These include the number of visitors returning to your site or the time spent watching your videos. You could also launch a mini-survey on your site that measures the willingness of your visitors to recommend your site to peers. It is also known as the Net Promoter Score (NPS).

**How to Optimize Your Website?**

While a detailed discussion on optimizing your website is beyond the scope of this book, here are twelve critical points to make your website more effective. Consider these, for example:

1. Before designing a website or any web project, ask yourself two questions—Why should people come to your website? Why should people return to your website?
2. What is the most important page of your website? Every page. Visitors can enter any page of your website through search engines and social media referrals. Create engaging content on every page.
3. Research suggests that less than 20 percent of visitors go to the home page. It is, therefore, essential to include a Call to Action (CTA) on every page. Remember, a webpage without a Call to Action (CTA) is a lost opportunity.
4. Physicians prefer independent, third-party websites to those pharmaceutical companies provide as they perceive them to be more trustworthy. It is, therefore, better for pharmaceutical companies to partner with independent websites to engage physicians more effectively.
5. Good design matters. Remember that your visual design will determine visitors' first impressions in many cases, no matter how valuable your content is. First impressions are crucial. A visitor decides whether to stay on your website or go to a competitor in a split second.
6. The entire site should scale to different devices. If the visitors find the site fiddly or annoying, they will leave regardless of its quality. It is also called *Responsive* design.
7. Monitoring your website results and adjusting when needed is crucial for success.
8. A/B Testing is essential. Try out different designs, wording, and flows. Changing the wording in a button can significantly impact conversion rates. Therefore, never stop testing.Testing the usability of a website is vital.

9. No matter how visually pleasant or content-rich a site is, users will get frustrated and abandon it if it does not work or is challenging to navigate. Customize your web pages. Adjust content to match visitors' needs. Develop multiple landing pages, depending on the nature of the visit.
10. Optimization of your website is vitally important. Your website should scale according to the device and screen being used. Ensuring every visitor browsing your website, irrespective of the device he uses, be it a smartphone, tablet, or laptop, has a great user experience. As Steve Jobs famously said: *Design is not just what it looks like and feels like. Design is how it works*. Combine content with services. You can build enduring customer relationships by combining your compelling content on your website with services such as continuing medical education (CME).
11. Continuous management of your website is not something nice to do but mandatory. Management of website, not just from a technical point of view but also concerning its content. Update your content regularly.
12. Finally, ensure that every page has credibility, trustworthiness, and professionalism. The tone of voice of your website should resonate consistently on every webpage.

A website is a vital tool for many companies, which has many purposes, including brand building, fostering customer loyalty, and 360-degree customer engagement. The following case shows what a well thought out website could do for the company.

CASE

15

# Dr. Reddy's Laboratories Create An Unbranded Russian Website on Pain Management!

Dr. Reddy's achieved a leadership position in the pain management area in the Russian market. The company garnered every second prescription written for alleviating pain in the market. The company's six brands enjoy the leadership position. Not content with this, Dr. Reddy's wanted to achieve the status of one of the most respected companies. To defend its therapeutic leadership in the pain management segment and stay at the top, the company conducted market research to identify the unmet needs of physicians in the Russian Federation. The company found that all the physicians wanted to acquire knowledge and update themselves about the latest medical developments related to their specialties in the Russian language because doctors in Russia study medicine in the local language, Russian, in the medical colleges across the country.

Dr. Reddy's created an unbranded portal on pain management to meet this unmet need. They chose an unbranded website because an unbranded website is seen as more credible by physicians. The company created compelling content about the latest pain management advances, thousands of articles, and key opinion leader (KOL) videos in Russian. The company updated the content at regular intervals, almost daily, and more than once a week without any breaks. The company sourced its content from the most respected scientific publishers worldwide. As a result, more than 18,000 physicians, six thousand of whom were neurologists, registered themselves on the website enabling the company to have a phenomenal reach of physicians that it could not cover.

The website was an unprecedented success and set a new benchmark. It achieved a Click-through-Rate (CTR) of 20 percent against the industry standard of about 3 percent. Even the email open rate was a phenomenal 6 percent, six times more than the industry's 1 percent.

Dr. Reddy's had also started closed-loop marketing with analytics at the back end and engaged physicians meaningfully with relevant messages that resonated with physicians. As a result, medical representatives of Dr. Reddy's could engage their physicians far more meaningfully.

The unbranded portal has significantly impacted its presence in the Russian market, reinforcing its strong leadership position.

## E-Mail Engagement

An email address or Digital ID is the most widely used form of digital verification. It can tell marketers about online people, where they have been, what they have done, and even what they like. This information enables marketers to improve personalization and deliverability and target their marketing messages to the right customers.

Customers today are more likely to respond to personalized interactions with brands and companies when and where they are most engaged. Today's technology uses customers' digital ID as the core of who they are, where they have been, what they like, and where they might go. When you consider the usage of emails today, you will understand how important they are as digital IDs. In 2021, the number of emails sent and received were a whopping 347.3 billion per day. Statista, the German company specializing in market and consumer data, estimates that the number of email users will go up to 4.6 billion by 2025.

An email has an ability that many other channels do not have. It can create valuable, personal touches at scale. Different types of emails are sent to different target groups through email marketing to build and maintain customer relationships. The three main types of emails are:

1. *Marketing Emails* such as newsletters, sales, and promotional announcements are commercial.
2. *Notification Emails* are more informative. These include registration confirmation emails, status update changes such as updates from free to premium services, notifications on recently browsed products, feedback request emails, and subscription abandonment.
3. *Transactional Emails* are sent confirming the customer relationship. These include welcome messages, account opening, shipment tracking and order status updates, and more. The recipients expect these.

Research indicates that about 60 percent of Pharma marketers consider email marketing a standard practice in their marketing activity. Best Practices research suggests that Pharma companies would do well with their email strategies if they do three basic things:

1. The subject line of your email determines whether your email will be opened or not. It is because your recipient's first impression of your email communication is through the subject line. Considering the daily emails we all receive, the subject line must be persuasive to generate an action. Research suggests that subject lines of under 49 characters make the best open rates. You must start your subject lines with information-carrying and descriptive words. Also, avoid generic, meaningless words in the subject line.
2. Make it easy to subscribe and unsubscribe from your emails. When you make it difficult to unsubscribe, people get annoyed and may even mark you as spam, which can be more harmful.
3. In the Pharma industry, open rates for rep-triggered (sent from sales representatives) emails can be twice as high as those for mass campaigns.

### 4-*U* Approach to Subject Lines

Subject lines are crucial for the success of email marketing campaigns as they determine your email open rates and click-through rates (CTRs). Therefore, you have to prepare your subject lines carefully. A 4-U approach is a useful tool in this regard. Rudd Kooi, Fonny Schenck, and Beverly Smet proposed this 4-U Approach in their insightful book, *Evidence-Based Multichannel: Delighting Pharma Customers in the Omnichannel Age*. Consider these 4-'U's, for example:

1. *Useful*: In what way is the message of value to the recipient? Follow the WIFR-Rule: What's-in-it-for-the recipient?
2. *Ultra-Specific*: Make it immediately clear to the recipient what he can expect from the email?
3. *Unique*: In what way is the message compelling and inviting?
4. *Urgent*: Can you create a need to read it now?

## Effective Email Engagement

The lifespan of an email is very short. Twenty-four percent of all email opens happen during the first hour after delivery. The chances of your email being opened after twenty-four hours are less than one percent. So how do you ensure that your audiences open your emails? How can you engage them effectively? Here are two factors that can help you significantly engage your audiences effectively.

1. *Personalize your messages*. The vital first step is knowing how, where, and when to personalize messages. As more and more people are worrying about cyber-security, personalization has to be used with caution. Trust is essential in personalization. Like in real life, you build a relationship of trust over time. Any insincere familiarity can put the readers off. Personalization should go beyond addressing the reader with classic Dear followed by first and last names. Personalizing content is more important. Use customer journey and information to suggest content, products, or services. Giving people what they want helps forge closer ties. What parts of the email can be personalized? Virtually every part! Consider these, for example, Subject lines, Content blocks, Offers, discounts and offers, specific information relevant to that consumer, CTA (Call-to-Action) links, or buttons.
2. *Communicate Segment-specific messages*. Research suggests different email behaviors between general practitioners (GPs) and Specialists. GPs usually generate a higher open rate of 27 percent than specialists, with 22 percent. While this may not be a big difference, it is significantly different for the click-through rates (CTRs). Specialists showed a 21 percent CTR on average, and GPs displayed a considerably lower 13 percent CTR. Therefore, you should design your emails using subject lines and segment-specific content for GPs and specialists. Additionally, it would be best if you keep the following factors in mind while designing segment-specific emails:

- You need to customize content for different personas of customer segments.

- For loyal customers, for example, content should be focused on expanding prescription use in a new group of patients.
- In Non-prescribers, the focus of content may be on overcoming the hurdles of prescribing for the first time.
- Also, you can send different content depending upon the orientation of the customers. You can design science-focused content for a more science-oriented specialist physician and a different type of content for a more relationship-oriented GP.

### Measure What Matters

The basic metrics that you should measure and monitor in your email strategy are:

1. **Exposure Metrics** look at exposure, or how many people you can reach with your message. Good exposure metrics are the number of opt-ins you collected, the number of emails sent, and the number of email permissions obtained over time.
2. **Interaction Metrics** look at interaction levels. Good interaction metrics are the number of emails opened, and emails bounced.
3. **Engagement Metrics** look at engagement levels. These are the most important as they measure the true impact of an email. Good engagement metrics are the number of click-throughs, or call-to-actions responded to, the number of opt-outs, referrals, and shares.

### A to Z of Email Marketing: Winner's Checklist

How to win with email marketing? What are the critical success factors? Here's a checklist of essential steps to pay attention to and act upon. Finally, here's a list that can help you win with your email marketing.

A. Who will receive your emails? Define your target audiences clearly.

B. Define relevant customer segments so that you can customize messages.

C. Define your email content. Prepare a detailed content strategy. Which type of email is to be sent to whom? Consider all kinds of emails, such as marketing, notifications, and transactional.

D. What are your frequency goals? How often do you plan to send your emails? What are your key performance indicators for each type of email?

E. Prepare a schedule covering all your email messages.

F. Use an actual person's name as the sender of an email. Having a generic 'info@' sender address can hurt open rates.

G. Have a clear, attention-grabbing subject line that tells the recipient exactly what they should expect in the email.

H. It is vital to have a personal salutation for each email.

I. Anchor your email with the value proposition mentioned in the subject line. It could be a downloadable report or an article.

J. Personalize your email offering to the extent possible.

K. Make your CTA (Call-to-Action) easy to recognize and understand so the recipient knows what she needs to do.

L. Ensure that all images for the mailing are relevant to the content and the subject because images will not be shown automatically in most email clients. People need to download them. So avoid putting your CTA (Call-to-Action) buttons in the image of an email.

M. Include social sharing buttons allowing recipients to share your message with their network in your email.

N. Prepare your designs and content for your emails. Test them and ensure that they work equally well on all devices. Optimize your emails for mobile devices. Over 61.9 percent of people open their emails on their mobiles today.

O. Now about your header and footer of the email. Ensure every email contains a link to your privacy policy, an unsubscribe option, and a link to your website. In addition, always provide an option to view the email on a webpage in the header of your email in case of recipients cannot open it in their email client.

P. Make sure your email message contains a sufficient amount of actionable links. More links mean more clicks.

Q. Ensure you receive all the emails sent so you can better understand how your target audience receives them.

R. Try to standardize your email templates as much as possible. It will save development time while at the same time will ensure a consistent look and feel.

S. New subscribers are often more engaged. Do keep track of the date when people subscribe to your program. It will help you adjust your content, as subscribers' interest may start to wane after a while.

T. Facilitate subscription. Make it easy for recipients to subscribe and unsubscribe.

U. Make it easy to share and forward your email content.

V. Email is an essential part of every brand team's communication mix. Therefore, this channel should be owned by the marketing, not the IT department.

W. Measure results. Track open rate, click-through rate (CTR), unsubscribe rate, bounce rate, website traffic, and signups since the previous campaign.

X. Continuously test different features of your email campaigns, such as design, different call-to-actions, or different wordings, as these things can significantly impact campaign outcomes.

Y. Continue to optimize audiences' content, design, and frequency.

Z. Finally, email is an essential part of every company's communication mix. Investing time, effort, and resources in email marketing is a *must-do* despite the advent of newer social media sharing channels.

## Mobile Engagement

Mobile is one of the most recent disruptions in the digital space today. Every digital activity you can think of, such as social sharing, searching the internet, buying things, and making payments, has gone mobile. As a result, mobile, one of the recent innovations in digital technologies, has enormous potential in almost every industry, including the pharmaceutical industry. However, mobile is not without challenges. Companies get only one chance to make a first impression, mostly via a mobile device. If your website is not mobile optimized and does not scale up properly to your customers' mobile devices, you are on a sure path to losing your customers. Therefore, it is important to follow a mobile-first approach.

### Mobile-First Approach

As the name suggests, the Mobile-First approach is about designing a desktop website starting with mobile and then scaling it up to larger screens. Thus, the mobile-first approach is contrary to the traditional approach of starting with a desktop website and then adapting it to the mobile. Instead, a mobile-first approach means building your website with mobile in mind, with the main objective of improving the mobile users' experience on your site.

The rapid adoption of mobiles makes it mandatory for companies to adopt a mobile strategy. Consider these facts, for example:

1. According to GSMA (Global System for Mobile communication Associations), over two-thirds (67 percent) of the world's population had a mobile device in 2021. In 2017, for example, a little over half (53 percent) of the world's population had mobile devices. Statista, the world's leading business data platform, predicts the number of mobile device users will increase to 7.33 by 2023 from 7.1 billion in 2021.
2. The rise of mobile devices has led to an *always-on* mentality, blurring the lines between private and professional lives.
3. Research indicates 9 out of 10 smartphone users have their mobile within arm's length round the clock.

4. Also, eighty-seven percent of smartphone users read personal and work-related emails on the same mobile device.

### How to Develop A Mobile-First Mindset

Here are five key questions you need to answer to develop a mobile-first mindset.

1. What is your level of mobile maturity? Are you a beginner, or already have taken some mobile initiatives? If you are not a novice, how many mobile initiatives do you currently have, and how well are they integrated into your overall marketing strategy?
2. How do you compare with your closest competitors on mobile maturity?
3. Is your website mobile responsive? If it is not, make it now. Then, after building a mobile-optimized website, you can start creating apps when there is a specific need.
4. Are you considering mobile as an *add-on* to your strategy or fully integrating it into your overall marketing strategy?
5. Are you capturing learnings from your mobile initiatives and building them into future strategies?

### Mobile in Pharma

Pharma companies need a clear mobile strategy and mindset to survive and grow in the rapidly changing marketplace. Therefore, mobile is must for Pharma. Consider, for example, the following reasons:

1. According to Statista, the number of mobile users globally is likely to reach 7.49 billion by 2025 from 7.26 billion today in 2022.
2. More than one-half of all online searches take place on mobile. Fifty percent of all web activities are mobile.
3. Fifty-two percent of smartphone users gather health-related information on their phones.
4. Even HCPs (Healthcare Professionals) are turning to mobile. About 80 percent of physicians use smartphones and medical apps.

5. In 2021, Health and Fitness Apps generated 2.5 billion downloads, which are well above pre-COVID levels.

In addition, most healthcare professionals actively search and consume professional content outside normal business hours. Therefore, pharmaceutical marketers should design mobile services with different user contexts and times.

When your target audience is on mobile more often, you cannot afford to ignore mobile. Going mobile is a way to engage them when they want it where they want it. It's no longer an option.

**State of Mobile Marketing**

While most pharma marketers acknowledge the importance of mobile, only seventeen percent of them have their mobile strategy fully integrated into their overall marketing strategy. And only half of them have a mobile-first mindset.

The variety of devices and operating systems make it difficult for marketers to embrace a mobile-first mindset. That is probably why many of them are stuck in a mobile-campaign mode, and very few reach a mature mobile digerati state. For example, Research2Guidance, which studied the pharma app field in 2017, found that Johnson & Johnson was the only pharmaceutical company with a good mix of breadth (number of apps) and depth (usage).

**COVID-19 Accelerates Mobile Health Growth**

The global COVID-19 pandemic has led to a significantly increased utilization of mobile health platforms across the globe. In addition, the pandemic made remote patient monitoring, patient data analysis, and diagnosis and remote engagement of pharma sales reps and physicians absolutely necessary.

Mobile health or medical application downloads during the peak month of the pandemic recorded an unprecedented increase. According to Statista, the leading business data platform, there was a 135 percent increase in medical application downloads in South Korea and around 65 percent globally during the peak month of

COVID-19. Thus, the pandemic has boosted mobile medical app adoption fueling its market growth.

### Mobile Marketing: Key Tactics

Mobile marketers use many tactics to reach and engage their customers to have a *must-have* status. Consider, for example, these tactics:

1. A responsive website or a mobile-optimized website has become the norm today. Websites that do not scale properly to whatever mobile device their customers use are bound to lose customers.
2. The second most important mobile tactic is a mobile-optimized landing page.
3. Outbound SMS (Short Messaging Service) is another popular tactic for using text for multimedia messages and notifications. Short-code SMS replying is a tactic that allows customers to reply to an SMS sent during a marketing campaign with a short code.
4. Mobile bar codes, such as QR codes, conveniently redirect customers from offline channels, such as brochures, to the online space. QR codes are graphics that can be scanned by a mobile device and then redirect customers directly to an URL.
5. Geo-targeting allows marketers to deliver specific and relevant content on a mobile device based on the website visitor's geographic location. For example, a website visitor can receive a targeting message about a flu outbreak when entering a geographic location with a flu outbreak.
6. Wearables are another very important mobile tactic. Wearables are mobile devices with sensors that track motion, heart rate, sleep, or other physiological metrics. You can find them on smartphones or dedicated devices such as wristbands, watches, and even shirts.

## Wearables and Disappearables

Wearables and devices worn on the body continually collect data such as biofeedback and physiological data. They then upload the data automatically to the cloud to make it available for review by HCPs or patients. This automatic upload allows for greater accuracy and consistency of health data collection. However, there is a long way to connect and integrate this data directly into a more holistic healthcare record system.

There is an increasing trend towards becoming the quantified self or life-logging group. There is a growing interest among people to keep track of their health records to adjust their behavior and stay healthy without the input of a physician.

There is another trend that is increasing. *Wearables* are given away to *disppearables*. Disppearables are biodegradable sensors inside the body. Some biopharma companies are taking to the internet of things in a big way and introducing disppearables as a solution around the pill.

## mHealth

mHealth is a term used for medical and public health practice supported by mobile devices. It refers to mobile communication devices such as mobile phones, PDAs, and tablets for health services and information. Mobile devices are changing how patients and healthcare professionals access the internet and find health information.

mHealth is helping track some healthcare challenges, such as the aging population, non-compliance (patients not taking medicines), management of chronic diseases, and even helping with strategies for the prevention of chronic diseases. Here is a partial list of areas where mobile device applications are used:

1. Reference
2. Decision Support
3. Monitoring
4. Compliance

5. Disease Management
6. Chronic Disease Prevention

Increasingly patients depend less on HCPs and more on online for healthcare information. It is, therefore, important that pharmaceutical companies and other healthcare organizations provide accurate information. In addition, pharmaceutical firms and healthcare marketers can use new technologies such as geofence locations, video assets, and text messaging to reach their audiences, even in doctors' offices or near a pharmacy.

Furthermore, mobile phone and tablet applications have provided a new and easy way for patients suffering from chronic diseases to manage their conditions and guide them toward healthier lifestyles.

**More Apps**

Pharmaceutical companies realized the importance of Gamification as another tool to engage their key stakeholders, such as patients and HCPs, and started developing more and more apps. Here are some examples of successful apps:

1. *Respimat inhalation demonstrator by Boehringer Ingelheim*. Designed for HCPs, the app provides medical information on the next-generation device and facilitates physicians training patients on using Respimat. The app enables inhaler interaction, a vital issue, offers hands-on experience with an intuitive interface, and is available at no additional cost.
2. *Back-in-Play by Pfizer*. Designed as a game, this app aims to boost knowledge of a little-known disease, ankylosing spondylitis. Through this, European patients learned about this condition that causes inflammation in their spine and pelvic joints.
3. *Boehringer Ingelheim's HealthSeeker* is a game that helps those with diabetes make better lifestyle choices about their condition and overall wellness.
4. *Bayer created the diabetes Didget* game using Nintendo's DS handheld game system to regularly encourage children with diabetes to test their blood glucose levels. Players receive

rewards for testing consistently. Bayer's *Didget* blood glucose meter fits with the Nintendo handheld game system. It allows children with diabetes to play Knock-'em-down, the World's Fair Board Game, inspiring pediatric patients to keep up with their testing schedule.

5. *Johnson & Johnson's Care4Toda*y is designed to improve adherence to treatment regimens through reminders to take medication, refill prescriptions, and visit healthcare providers. The mobile adherence program can be used for any prescription or OTC medication or nutritional supplement, including but not limited to the Johnson & Johnson group of companies.
6. *Janssen's 'Sorted App'* is designed for patients with attention deficit hyperactivity disorder (ADHD) in the UK. The app acts as a daily organizer to help patients create, categorize, and prioritize daily tasks with a point-scoring element to motivate the users. Janssen's key product for ADHD is Concerta XL. Developed with the concurrence of Physicians in the UK, the app is designed to utilize the functionalities of the iPhone, such as its calendar, reminders, and voice memos. The app's targeted at people above 12 years.
7. *Sun Pharma's RespiTrack App* is a mobile application connecting doctors with asthma patients. The app focuses on improving treatment compliance and minimizing non-adherence and dropouts in the asthma treatment regime. RespiTrack allows complete monitoring of asthma attacks, symptoms, and medications prescribed by physicians. It enables patients to record the time, place, duration, and triggers of attacks along with medication details on the mobile app. The patient can easily send this data to his physician through over-the-air mode weekly or monthly. In addition, a review dashboard in the app provides a snapshot of all parameters for monitoring periodically. Also, through RespiTrack, a doctor can access his patient's details through a personalized tracker module, which is directly synced with the patient's app. This module allows doctors to access patients' ailment history, attack patterns, symptoms,

medication patterns, and reports. In addition, a doctor can remotely view patient reports and connect with him through an integrated messaging app.

The following two cases show how mobile apps can help pharma marketing :

- Case of Alkem, an international generic drug firm from India, used a mobile app to help patients improve their medication adherence and health monitoring, and
- How *One Drop*, a mobile app for monitoring and managing diabetes, achieved significant success.

CASE

16

# Alkem's Initiative into Digital Marketing with *DonApp* is Successful!

Alkem is the latest example of what digital marketing can do for your brand's success. Digital marketing can do a lot for your brand. For example, digital marketing has helped Alkem to grow its DonApp brand's market share from 37 percent to 43 percent in one year.

DonApp is the Alkem brand of Donepezil, a drug that effectively treats dementia and Alzheimer's. Donepezil is originally discovered by Eisai Pharmaceuticals, a Tokyo-based international drug company. So naturally, Eisai was enjoying brand leadership in the Donepezil market with its brand Aricept.

Alkem started its digital marketing foray in 2016 and never looked back. The company launched a patient-focused digital campaign in its marketing, a departure from the doctor-focused industry norm. In addition to its conventional transactional mode of marketing, the company did four things in the digital space.

First, it formed a 'Dementia and Alzheimer's Community of India' Facebook group with 75,838 members as of September 18, 2018.

1. It started sharing emotional videos on WhatsApp.
2. In early 2016, Alkem launched an *I Remember* campaign that helped achieve its dual objectives. The company spread awareness about dementia, a mental condition that people are hesitant to talk about, and allowed it to lead a growing prescription drug segment.

3. Alkem launched a mobile app with its brand name - DonApp, specially made for Alzheimer's' patients and their caregivers. It can give your loved elders helpful reminders about medicines, food, doctor appointments, and a location tracker.

Alkem's digital marketing initiatives are noteworthy because only 7 percent of Indian drug manufacturers delve into digital marketing. Whereas for multinational companies operating in India, it is becoming a way of life as their respective parent companies have been practicing digital marketing for some time. Indian drug laws, however, do not allow a pharmaceutical manufacturer to promote a prescription drug to the public. However, it does not stop the company from helping dementia and Alzheimer's patients with GPS-tracker-enabled free apps if patients go missing. That's why Alkem chose to launch a mobile app with its brand name - DonApp, for both IOS and Android mobile users.

More than 4.1 million Indians suffer from dementia, the most common being Alzheimer's disease, which causes a gradual decrease in remembering and thinking clearly, eventually affecting a person's daily functioning. However, the awareness of the disease and its consequences is low due to social stigma and lack of awareness, and people often confuse forgetfulness associated with normal aging. Alkem is also partnering with the Alzheimer's and Related Disorders Society of India (ARDSI), Mumbai Chapter, to spread awareness. The society conducted a three-kilometer Awarathon on December 3, 2017, at Mumbai's Worli sea face.

The potential shift of power from doctor to patient is evident from the Government's determination to stop the physician-pharma nexus by legalizing the Uniform Code of Pharmaceutical Marketing Practices (UCPMP) and the internet explosion bringing all information that was privy only to the healthcare professionals till now easily accessible to the public. Thus, the pharmaceutical industry in India and the world over have an excellent opportunity to share medical information with patients through social media channels and educate the patients.

Pharmaceutical companies in India, even today, rely heavily on transactional marketing and are following the conventional methods of engaging physicians with high-value gifts, sponsorships and vacation packages, cruises, and even cash payments. For all such pharmaceutical companies, digital marketing could be a rite of passage into a patient-centric, information-driven, and corporate-brand-building-focused approach from the conventional *quid pro quo* relationship with the doctors.

While this path is not easy, Alkem has shown it is possible. Alkem grew its DonApp brand's sales from Rs. 23 crores in 2016 to Rs. 26 crores in 2017. It may not be much in value, but the unit-wise sales increase has been impressive because the drug's price was capped in March 2016 by the regulator. Much of this growth can be attributed to its social media campaign.

Alkem's digital marketing initiative certainly paid rich dividends in every way you look. It indeed shows what can be done with digital marketing.

(Source: Adapted from Ruhi Kandhari's article, *Indian Pharma doesn't have a choice but to market digitally*, The Ken, November 1, 2017)

CASE

17

# Mobile App Success Story: How *One Drop* Did It!

How did *One Drop* become so successful as an app for monitoring and managing diabetes? Because it offered solutions that match users' needs by paying close attention. It is built to help people who struggle with diabetes every day. It was created by Jeff Dachis, who was a diabetic himself.

Jeff Dachis discovered in September 2013 that he had diabetes. He started to search for solutions for managing his condition but found many gaps in the medical system. Finally, he understood how scary it could be for a person to be diagnosed with diabetes and clueless about it, and that is when he decided to build his app to help all patients with type 1 and type 2 diabetes log the required data to improve their health. That's how the mobile app, One Drop, for managing and monitoring diabetes effectively, was born.

Jeff Dachis launched the *One Drop* for iPhone mobile app in April 2015 on Apple App Stores, after a year and a half for Android users. Today, it continues to be a trustful partner for those who need a great amount of information about type 1 and type 2 diabetes.

## What Makes *One Drop* Unique?

What makes *One Drop* unique? There are at least four things that come to mind. *One Drop's* success rests on these four pillars: Simplicity, Community, Technology, and Business Model. These factors have not only made *One Drop* very successful, but they have also been responsible for winning the best design award at its launch festival in 2015, one month before it hit the market. After that, every important publication, such as TechCrunch, Mashable, CNN, Diabetes Daily, Fortune, the Washington Post, Gizmodo, and many others, revealed the impact of this app on its customers. So what made One Drop the big success it is today? Here are four major factors that made One Drop a big success today.

1. *Simplicity*: When dealing with massive data, providing the right information at the right time becomes crucial for making your life easier. Management of diabetes involves the processing of a lot of data. Data include your dosing schedules, monitoring of blood sugar levels, and nutritional and caloric values of the foods you take, your walking and exercise moments, among others. With One Drop, you can schedule specific reminders for medication, and you will receive insulin pump data. The same app will calculate the carbs for each recipe, giving you a better overview of your progress.
2. *Community*: As Jeff Dachis, the CEO of One Drop, says: *Diabetes is hard; it doesn't have to be*. It is especially true when a great group of people who share the same concern allows you to check their evolution of progress, recipes, and other information that becomes useful to you if you are determined to live a healthy life. Their social support can be of immense help, even when you are not confident about your progress. Interacting with other members of the same community makes you feel a little better. Also, what others accomplish makes you see that you can reach the same level. What is more, the news feed of the One Drop app keeps you abreast of all the developments and events in the area of diabetes. The app helps you build lifelong relationships with the community.
3. *Technology*: One Drop is more than just an app for diabetes management. It helps all categories of people who are interested in managing their overall health. For example, if you

usually forget to take your meds, no matter what reason, this is an app to use. At the same time, if you want to keep your weight in check, One Drop is a suitable solution. Also, the app's smart tools help you with accurate data to help you eat right and exercise appropriately. One Drop is an app that allows users to maintain a positive lifestyle and attitude. *Furthermore, One Drop has recently added an Augmented Reality (AR) toolkit* that presents advanced data visualization technology to help users to analyze their graphs better. An engaging interface offers a pleasant experience while customers understand the stats displayed. One more recent additional feature of One Drop to take its technology to a higher level is its integration of the blood sugar monitoring system with Apple Watch and an Automated Decision Support System. One Drop became the first and only wireless blood sugar glucose monitoring system to connect directly to Apple Watch. The app allows users to transfer blood glucose data from One Drop seamlessly. OneDropChrome transfers the blood glucose data directly to the Apple Watch companion app. With its new Automated Decision Support system, One Drop forecasts blood glucose prediction values over time and offers behavioral recommendations. One Drop's proprietary Artificial Intelligence ( AI) generates accurate predictions for a single user based on over 1.3 billion aggregate longitudinal Diabetes health data of all people utilizing the One Drop platform. Uniquely, this allows One Drop to provide a user with accurate predictions after entering a single data point and improves over time.

4. *Subscription Model*: While the basic app of One Drop is free to its users on Apple App Stores and Google Play, there is a subscription model that offers many value-added services. The basic app provides everything people with diabetes need to manage the data required for healthy living. For those who need more and require 24/7 assistance, for a monthly subscription, they receive:

A. Personalized help, and they will be coached by a certified diabetes educator(CDE). It is essential to count on specialized support whenever you need it. With One Drop —Experts, users have two options: on-track and on-call. Whatever their choice, experts will be there to encourage and help them continue the program.

B. Additionally to this in-app chat feature, users who pay for extra services will receive test strips, and those who will take a One Drop Chrome, a blood glucose meter that uses Bluetooth to sync data with the One Drop mobile app.

It is tough to control chronic conditions such as diabetes, but with the new technology that appears daily, everything gets easier for patients and their families. One Drop has become one of the most robust solutions for diabetes, weight loss, hypertension, and diabetes prevention. As one of its customers described: *One Drop* can be a lifesaver!

Speaking about the empowering nature of their latest application of proprietary technology of Artificial Intelligence and Augmented Reality to their already famous diabetes management system, One Drop, Jeff Dachis, CEO of One Drop, said: *For too long a time, the diabetes industry has operated on point solutions and closed systems focused on the needs of healthcare providers in the clinic; rather than on the needs of the people using the systems to manage their disease every day. We are excited to further empower One Drop users with their addition to the* One Drop *experience*.

(Source: Adapted from articles: 1. Felicia, Mobile App Success: How One Drop did It, App Samurai, October 2, 2017, 2. News Provided by One Drop, One Drop became the first diabetes glucose monitoring platform to integrate directly with Apple Watch, CISION PR Newswire, September 12, 2016)

# Gamification

You can find Gamification everywhere today — from boardrooms to classrooms and social media. Over the last few years, Gamification has gained wide acceptance and recognition in many areas, such as marketing, healthcare, business, politics, and technology design. The pharmaceutical industry has embraced this trend in healthcare gamification to meet the ever-increasing challenges of creating value and brand building. Initial research and case studies prove that Gamification improves patient compliance and health outcomes.

### Five Key Elements

Gamification uses gaming mechanics to engage users and solve problems in a non-game context. There are five critical elements in a gamification strategy:

1. Reward or recognition for the players' achievements.
2. A goal and a sense of purpose.
3. Rules and a framework for participation motivate participants to tackle challenges strategically.
4. Continuous challenge. Achieving each next step should be harder but not too hard. The best option is that level one prepares the user for level two. There should be a flow that will keep the users fully engaged.
5. The feedback that tracks the players' status informs them how close they are to the end goal.

### How Gamification is Beneficial

1. Gamification can promote therapies. By creating interactive, educational health content, Gamification can promote new and existing therapies.
2. Gamification can increase engagement between Pharma, HCPs, and patients. HCPs' use of interactive technology to

communicate with their patients is gradually growing. With Gamification, Pharma can provide responsible, educational health content and improve health awareness among patients. Sanofi-Aventis has successfully used Gamification as a strategy to improve patient adherence.

3. Gamification allows showing patients firsthand how their daily habits and choices affect their condition.
4. Gamification helps create and regulate virtual patients by inserting medical record information.
5. Gamification can facilitate physician education. Physicians are interested in improving the health outcomes of their patients in less time. Pharma companies can help achieve this by implementing Gamification.

**Slow Rate of Adoption**

A research study by Research2Guidance (R2G) observed that between 2008 and 2016, the top Pharma companies had 65 apps in the Apple and Google Play app stores on average compared to 1 or 2 apps from other health app publishers. However, even the Pharma companies with the most downloaded apps have accrued 6.6 million downloads since 2008 and had less than a million active users. So why is the adoption rate slower in the case of Pharma compared to other healthcare and fitness apps?

If Gamification offers so many benefits, why are there relatively fewer downloads for Pharma apps compared to other healthcare apps? Here are the possible reasons:

- Many of the Pharma apps have highly targeted audiences like HCPs.
- People with specific disease states.
- Pharma apps tend to target local markets in three or fewer countries, making them unlikely to compete on download numbers with fitness, health, or diet tracking apps that appeal to a larger market segment.

**Engagification, not Gamification!**

While Gamification positively impacts participant engagement, message recall, and patient outcomes, one problem exists. It is the name itself Gamification. When you say that you will gamify the participants' interaction with your content, the term, Gamification lends itself, to trivializing the most serious experience related to improving health and well-being. Patients, physicians, and other pharmaceutical industry stakeholders still cling to the idea that games should not be part of the conversation in such a serious realm as healthcare. In Pharma marketing, it is crucial to steer the discussion away from the word Game—and towards terms like engagement, retention, and adherence. Therefore, please do not call it a game. Call it an app. More importantly, your strategy is not Gamification. It is engagification!

## Multi and Omni Channel Engagement

As the name indicates, multichannel marketing (MCM) uses multiple channels to reach a target audience. In the case of pharmaceutical marketing, the target audience is physicians, followed by healthcare professionals, patients, and other stakeholders. The communication channels include both traditional and digital ones. Pharmaceutical marketers, however, have always operated in a multichannel environment that included personal selling, direct mail, journal advertising, meetings, and conferences, even before the internet era. However, these channels tended to be marginal and not necessarily integrated with the dominant channel, the field force.

Multichannel marketing involves multiple channels — offline, online, personal, and non-personal. As the internet has grown exponentially, the number of channels too increased dramatically. The following table presents personal and non-personal pathways in the pharmaceutical industry.

**Table 6.2 Customer Engagement Model in the Pharmaceutical Industry**

| Personal Interactions | Non-personal Interactions |
|---|---|
| Tablet Detailing | Website |
| Face-to-Face Meetings | Email |
| Local Physician Meetings | Direct Mail |
| Call Center Service to Customers | Short Messaging Service (SMS) |
| Medical Science Liaison (MSL) | Mobile Apps |
| Medical Conferences | Webinars |
| | Virtual Detailing |
| | Display Advertising |
| | Social Media Platforms |

One of Pharma's main challenges is selecting the right set of channels and, more importantly, the right content for these channels per segment or even at the individual level. Some channels, such as remote detailing, and tablet detailing, are more personal, but most can be considered non-personal.

There is more to multichannel marketing than multiple channels. It is essential to understand the difference between multiple channels and multichannel. Multichannel means getting synergies from connected channels. Simply adding channels without integrating them doesn't bring additional benefits. Your digital channels should be a connected network rather than a series of isolated silos of information. One channel should lead to the next.

According to Healthcare Dimensions, a firm specializing in Health Management Solutions, about seventy-five percent of Healthcare Professionals (HCPs) want to receive information about continuing medical education opportunities from pharmaceutical companies in an easily accessible manner and on demand. Integrated multiple-channel communication is the way to do that. Multichannel marketing offers three significant advantages over traditional ones.

1. *Relevant Communication*: Different people want to consume content in different ways. Advanced multichannel and omnichannel communication enables this by providing flexibility to deliver your content in the channel that best suits your audience's needs. The data gathered from the interactions with the tablet detailing or e-detailing presentation will inform the microsite content for each doctor. In effect, each doctor can now have their tailor-made channel and immediately connect to it. Pharma marketers, thus, can provide a better experience for HCPs.
2. *Better Clinician Education*: Multichannel communication offers another significant advantage. It enables you to demonstrate value with digital clinician education. The flexibility of multichannel allows HCPs to benefit from on-demand educational content and guided e-learning tools. As a result, the pharmaceutical industry can better understand its products and services by providing relevant and high-value information. It will help them raise their reputation and earn much-needed trust.
3. *Improves Treatment Outcomes*: Besides clinical education, you can use multichannel communication to help physicians to support patients throughout their healthcare journey by assisting them in improving adherence to medications.

## Benefits for Pharma

Pharmaceutical companies, too, can benefit significantly from multichannel marketing. Consider these advantages, for example.

1. *Increasing Marketing Return on Investment (ROI)*: Data drives multichannel marketing. It allows you to measure your marketing materials and efforts' return on investment (ROI). Because the channels are connected, you can also connect the data with a complete overview of what, when, and how content is accessed. You can see what is working and act quickly to make improvements.
2. *Empowers Field Force*: Multichannel empowers the customer-service teams in the field. Your medical representatives can supplement their face-to-face meetings with physicians with online remote contacts such as virtual detailing, eMSL (electronic-Medical Science Liaison) meetings, and other forms of virtual engagement. Besides, medical representatives can help physicians access different channels with on-demand content and offer personalized content to suit their individual needs.
3. *Refining Segmentation*: Many pharmaceutical companies recognize the need to refine their segmentation strategies by targeting high-priority customers with smaller field forces. Multichannel enables a key account strategy by delivering customer experiences that match your refined market segmentation. You can, for example, use high-value, one-to-one channels with priority targets by supplementing them with on-demand information relevant to their needs and wants.

## Common Pitfalls

Many misconceptions about multichannel marketing can prevent you from achieving your objectives. One should be aware of them and develop a more precise understanding of what multichannel marketing is and what it is not. Here are some common pitfalls you need to be aware of.

A. Multichannel, as such, is not about adding digital channels to your marketing mix. Some offline channels continue to do well and should be given due importance.

B. Digital is a strategy, first and foremost. A failure to see digital as a strategic investment will lead to failure.

C. Digital is often associated with innovation. When we do something digital, we do something innovative. No, it is not necessarily so. Digital is rapidly becoming the way we do business.

D. Many marketers seem to see digital as a channel in isolation. You need to activate and integrate all the channels — online and offline, along the customer journey, in a well-orchestrated way to achieve success.

E. Pharmaceutical marketers often see digital as a way for their field force to enhance engagement with physicians. However, such a stance is very limited. Digital offers a transformational opportunity to engage healthcare practitioners (HCPs) and all the new stakeholders. They are becoming increasingly important in healthcare decision-making, such as patients, caregivers, nurses, and pharmacists. Digital helps you to reach audiences that previously were not reached effectively through traditional and costly Pharma channels. It helps you to reach all the stakeholders cost-effectively.

## Winner's Checklist

How to succeed with multichannel marketing? What are the critical success factors? Here's a checklist for winning that can help you win with multichannel marketing.

1. *Determine your target audience*: Who are your target audience? Do you want to target patients, physicians, or payers? You can target as many audience segments as you want as long you have the necessary logistics to cover their entire customer journey.
2. *Measure your multichannel marketing success*: You cannot improve what you cannot measure. Set two or three key

performance indicators (KPIs) and a measurable objective for each channel. Track the progress against the set objectives. You must set goals to measure your reach, acquisition, and conversion. Monitoring the progress and making the necessary course corrections along the way is the key to achieving success.

3. *Integrate all your channels*: Integration of all your channels is vital for success. Channels need to connect naturally and not compete. Consider, for example, your blog post leads to a subscription box, a newsletter, and a landing page. Remember that your message is uniform across all channels. It is crucial.
4. *Engage with your stakeholders*: Customers are used to instant engagement and personalized messaging and content in today's Attention Economy. Your Pharma brand, therefore, needs to have a conversation online and not just a static presence. You can use live chat, mobile apps, email, chatbots, and more to engage with your customers with messaging that is as personalized as possible.
5. *Involve your representatives*: Pharmaceutical sales representatives continue to be a significant driving force in multichannel pharma marketing even though their numbers in the marketplace are dwindling. As such, you need to involve them in all aspects of the marketing process. Remember that emails sent by your representatives (Rep-triggered emails) have much higher open rates than those sent by pharmaceutical companies.
6. *Focus on the patient*: Patient-centricity is no longer a buzzword. Patients are increasingly playing an influential role in healthcare, and pharmaceutical marketers need to acknowledge this.
7. *Stay up-to-date*: There were few surprises with traditional marketing channels. It is not the case with multichannel marketing. Digital marketing changes rapidly, and you can sometimes be out of the loop in just a few months. Pharmaceutical marketers must stay updated with global marketing trends and not just stick to their industry. Moreover, you can adapt to change if you stay up-to-date with market

and marketing trends. Today, in the digital era, you can get data and learn from it almost instantaneously. The key questions to ask and answer are:

- Are physicians not engaging with your website enough?
- Don't patients find the information on your website useful?
- You can learn from the data you have at your fingertips and tweak your marketing strategy.

## Omni-Channel Marketing

The dictionary defines Omni as universal or of all things, whereas Multi means many. Therefore, the literal translation of Omni-channel means communicating across all channels rather than just many channels, as in the case of Multichannel.

What is the difference between omnichannel and multichannel marketing? The difference between omnichannel and multichannel comes from a company's digital channel approach. Stacy Schwartz, a digital marketing expert, puts the difference between omnichannel and multichannel marketing succinctly, when she says: *Companies that focus on maximizing the performance of each channel— physical, phone, web, mobile - have a multichannel strategy. They will likely structure their organizations into swim lanes focused on a channel, each with its reporting structure and revenue goals*.

The result is a competition that sometimes creates team friction and misalignment. But, on the other hand, sometimes it may work for the good. In multichannel marketing, channels are managed and optimized independently. It may result in competition within the marketing department or lead to a disconnected marketing experience. In contrast, *an omnichannel marketing strategy puts the customer, not the corporate silos, at its center*. It is thus more about customer centricity than anything else. The critical differences between multichannel and omnichannel marketing are presented in the table below.

**Table 6.3 Multichannel and Omni-channel Marketing: Key Differences**

| Multichannel Marketing | Omni-channel Marketing |
|---|---|
| 1. Brand is at the center of the strategy. Starts with the brand's view of the world and is then projected on the customer. | 1. Customer is at the center of the strategy. Considers how the customer sees the world and then determines internal marketing. |
| 2. Basically a tactical approach. Marketers manage each channel independently and implement tactics that work best for that channel. | 2. Internal departments work together and are in-tune with the overall messaging and strategy. The focus is on creating a consistent, cohesive overall strategy. |
| 3. Multichannel marketing is a brand's effort to interact with customers across multiple channels and platforms. | 3. Omni channel marketing offers a seamless, integrated customer experience, no matter where or how the individual customer engages with the brand. That's why it is called *Omni-channel.* |

Omnichannel marketing uses data across all touchpoints to build a customer profile to enable contextual marketing. Contextual marketing is an online marketing model in which people are served with targeted advertising based on terms they search for, or their recent browsing behavior. In the context of omnichannel marketing in the pharmaceutical industry, this means that if a physician seeks clarification or some additional evidence in a face-to-face detailing session with a Pharma repp and when he next visits an e-detailing site of that company, he should continue the journey, and not start from the beginning. That is a seamless customer experience.

It is important to remember that whether it is multichannel or omnichannel marketing, you should consider using traditional and digital channels in a mutually reinforcing way. Today there are many channels to reach the customer. The multiple channels that have cropped up in the digital era are the website, email, SMS (Short Message Service), Search Engine Marketing, Pay-per-click advertising, eDetailing, online forum, blogs, microblogging, social networks, big data and data analytics, and mobile applications.

### Winning with Omnichannel: A Checklist

How do you improve your chances of success and enhance omnichannel customer engagement? Here's a list of questions showing how to formulate a winning omnichannel strategy for engaging your stakeholders effectively.

1. What is your current understanding of the channels your customers are using?
2. What are the channels' reach, impact, and cost?
3. Do you have clear, measurable objectives for each channel?
4. What is the competitive landscape? How are your major competitors engaging their customers on these channels?
5. Have you looked beyond your traditional customer groups in your current strategy? Have you considered the increasing influence of other stakeholders?
6. How deep is your current customer segmentation? Have you considered the power of the finer segmentation capabilities of

digital marketing? Have you developed strategies based on personalization technology, or are you still focusing on mass outreach?

7. Have you put strategy first and channels second? Also, have you put content before channels? Do you have uniform communication that is consistent across all channels?
8. Have you considered the cost per engagement across all digital and traditional channels?
9. What is the level of integration of your customer-facing organization (sales, marketing, medical affairs, digital)? Can it provide a seamless experience for your customers?
10. Do you have an effective monitoring mechanism to track progress against objectives for each channel?

## Social Media Engagement

The times are changing for the healthcare environment in general, and pharmaceutical marketing in particular.

Like all other media, pharmaceutical marketing media was unidirectional and relied only on broadcasting its carefully crafted marketing message to its target audience, mostly physicians and doctors with good prescribing potential. However, since the mid-1990s, the growing popularity of the internet has offered a vast choice and freedom of access to content dynamically from millions of websites. Content generation during these early years of the internet revolution was no doubt massive, but there was not much difference in the type of content. Most of them resembled their traditional websites, which were more like online brochures, and e-commerce sites were merely looking like online versions of their earlier catalogs. This period was later termed web 1.0.

Later, Web 2.0 or Social Media emerged as a buzzword with the advent of user-generated content such as blogs and wikis. People started participating actively by using internet discussion boards, rating the content, and posting comments, making it a hub for social communities. What is most interesting about web 2.0 is that some of the most consumed and engaging information was not from a small number of media channels but from many regular internet users. As a result, early social networks emerged, enabling people to connect more easily. These networks matured and gave rise to services like Twitter and Facebook. Facebook today dominates the social media scene with a whopping two billion users.

### What is Social Media?

Social media generally refers to Internet-based tools that allow individuals and communities to gather and share information, ideas, personal messages, images, and other content. It also facilitates real-time collaboration with other users. Social media is referred to as Web 2.0 and Social networking. Social media includes blogs, social networks, video and photo-sharing sites, wikis, and many other media. An example shows some of the media grouped by purpose and functions. The list is only indicative and not exhaustive by any means:

1. Social networking (Facebook, MySpace, Google Plus, Twitter, and others)
2. Professional Networking (LinkedIn)
3. Physician Social Networks (Sermo, Quantia, MD, Doximity, Docplexus)
4. Media sharing (YouTube, Flickr)
5. Content production (blogs, Blogger, WordPress, Tumblr, and Microblogs such as Twitter)
6. Knowledge / Information aggregation (Wikipedia)
7. Virtual reality and gaming environments (Second Life)

### Social Media Marketing and Pharma

Social media marketing in the pharmaceutical industry combines various social media tools and platforms to establish conversations with physicians, patients, and other stakeholders. These online conversations help pharmaceutical companies build relationships with customers as you can spread brand messages through electronic word-of-mouth.

While traditional marketing media was more of a monologue, social media has created an opportunity for dialogue and conversation with consumers. Consumers can express their questions and concerns, enabling the brand to build a relationship. Social media users, in the process, generate a lot of content. They post over 30 billion comments on Facebook and more than two billion tweets on Twitter monthly. Moreover, about twenty percent of this vast generated content mentions a specific drug or disease. When one in every five conversation pieces is in the pharmaceutical and healthcare space, can Pharma afford to remain a silent spectator?

### Physicians on Social Media

While the exact numbers vary, most studies indicate that about 80 percent of physicians use social media for personal interactions, professional communication, and research. Physicians' social media usage seems to follow a typical 1-9-90 pattern.

- Approximately 1 percent of physicians using social media are content producers creating and publishing original content to other healthcare professionals and ePatients. They provide this information on blogs, forums, and information-sharing websites.
- A further 9 percent of physicians engage with others on social media by commenting on posts, participating in online discussions and chats, sharing useful information, and linking to other members online.
- The remaining 90 percent of physicians consume information. They use the internet and social media to find and read relevant information about their patients and practice.

### Patients Are in Control

More and more patients are turning to sites such as *Patients Like Me* and other patient and disease communities online to research their symptoms and arrive at a self-diagnosis before visiting their physicians. The availability of extensive healthcare information on the social web impacts the relationship between patients, physicians, and brands. Increasingly, doctors face self-diagnosing patients who have already researched their symptoms and are seeking prescriptions for conditions they think they have. The patients may even post reviews of their experiences in case of pushback by their doctors. Patients today are taking greater responsibility than ever for their health.

### Insights From Social Media Conversations

Patients seem more forthright and descriptive in their online conversations about their disease conditions and treatment feedback as they afford greater privacy than in-person consults. These conversations benefit pharma marketers as they help understand better how patients perceive the current product or service offerings and their unmet needs. With insights from this deep understanding, marketers can develop better, more relevant marketing materials, including packaging and educational and promotional tools aimed at physicians and patients.

## Challenging Regulatory Environment

Pharmaceutical companies, as they operate in a highly regulated industry, face many challenges, such as handling incomplete and misinformation about their products offline and online. According to the US Federal Drug Administration (FDA) Guidance on how Pharma should use social media, pharmaceutical companies should provide fair and balanced information about their products in their social media posts. In addition, they need to include the side effects, adverse reactions, and contraindications in their product information.

Furthermore, the FDA clearly stated that it wouldn't be taking a lenient approach towards either tweets or sponsored links. Social media posts must incorporate full and balanced information about their products, even on sites with limited character spaces like Twitter. It is not enough to provide links in the post to your web pages for detailed product information. Pharmaceutical companies face severe penalties from the FDA if they don't disclose properly. Drug firms must be careful not to mislead consumers on Twitter, Facebook, or any other social media channel. Otherwise, they will open themselves to liability. Kim Kardashian's Instagram post is a classic example of what happens when you don't follow social media regulations. Here is one case highlighting the importance of regulatory compliance.

CASE 18

# How Kim Kardashian's Instagram Post Got the Drug Firm that Makes *Diclegis* into trouble with the FDA!

In July 2015, Kim Kardashian, a leading supermodel with a massive following on social media, posted to her 45 million followers on Instagram and 35 million followers on Twitter about Diclegis, a prescription drug brand of a morning sickness tablet manufactured by Rosemont-based Pharma company, Duchesnay in the US. Here's a transcript of that post:

*OMG. Have you heard about this? As you guys know, my #morning sickness has been pretty bad...I tried changing my lifestyle, like my diet, but nothing helped, so I talked to my doctor. He prescribed me #Diclegis, and I felt a lot better; most importantly, it's been studied, and there was no increased risk to the baby. I'm so excited and happy with my results that I'm partnering with Duchesnay USA to raise awareness about morning sickness. If you have morning sickness, be safe and sure to ask your doctor about the pill with the pregnant woman on it and find out more:*

*www.diclegis.com; www.diclegisImportantSafetyInformation.com*

It is not a typical post by an average morning sickness sufferer who recommends a product online. Kim Kardashian is a celebrity with a massive online following and is paid to endorse the brand. The FDA took objection to this post as it failed to mention important information like the drug's side effects, adverse reactions, and contraindications. The social media post is false and misleading in presenting efficacy claims for 'Diclegis' but fails to communicate any risk information. The FDA demanded that Duchesnay immediately cease its 'misbranding' of the drug and has given time until August 21 to respond to the Agency. The company withdrew the social media posts. Here's what Dean Hopkins, general manager of Duchesnay, said: *In the original post we developed with Kim, we provided her with a link to risk information and limitation of use for Diclegis. But the post did not meet the FDA requirements for communicating important product information*.

Kim Kardashian posted a long corrective ad on August 30, 2015.

(Source: Sharon Lurye, How Kim Kardashian's Instagram Post Got Rosemont Drug Company in Trouble with the FDA, PhillyVoice, September 1, 2015)

### FDA Guidance on Social Media: Implications for Pharma

Pharmaceutical companies won't be able to send a series of tweets containing a product's full benefit and risk information. Instead, they will need to provide the complete information in a single tweet.

### How Can Pharma Comply with the FDA Guidance on Social Media?

Pharmaceutical companies must connect with patients and physicians on social media to offer reliable information about their products and improve health and well-being. However, how can they comply with the FDA regulations while providing this information, controlling the conversations on various social media platforms, and ensure fair and balanced information? Henniger Bullock and Collen Tracy Jones suggest a five-way approach to achieve this in their Law 360 article of April 21, 2017, *5 Social Media Pitfalls in the Pharmaceutical Industry*.

1. *Set A Clear Internal Policy*: Develop a clear policy that includes a centralized social media hub such as a company-sponsored website, Twitter handle, and Facebook page for posting messages on behalf of the company. Create a concrete social media advertising policy and clearly define who may engage on social media on the company's behalf. Ensure that all employees understand their role on social media, whether in their personal or official capacity, and how they must conduct themselves on social media sites. The legal department must clear the content for regulatory compliance.
2. *Create A Controlled Environment*: When consumers (patients) and customers (HCPs) are frequently on social media, can Pharma afford to shy away? No, certainly not. Pharma should embrace social media but maintain control of it. A pharmaceutical company can create a controlled environment, foster patient relationships, and communicate useful, fair, and balanced information about its products and services by creating a company website or a company-initiated chat area.
3. *Create A Compliance Strategy*: It is not enough if you define the roles of engagement and rules of conduct on social media

for your patients in your social media policy. You need to ensure compliance. A compliance strategy and a monitoring system are implemented to ensure total compliance.

4. *Comply With All HIPAA Act Privacy Laws*: Pharmaceutical companies need to do more than comply with the FDA Guidance on social media. They must comply with HIPAA (Health Insurance, Portability, and Accountability Act). Complying with patient privacy laws is equally important. Therefore, educate all employees on patient privacy laws and how they relate to business. Remember, even an individual posting a patient's picture violates privacy. Create a social media working group to discuss any concerns and issues.
5. *Keep Future Litigation in Mind*: Ensure you prepare a carefully coordinated message and approve it. Ensure compliance with social media standards and policies before embarking on a social media campaign.

### Pharma Must Embrace Social Media

Although many regulations regarding how pharmaceutical companies should conduct themselves on social media are challenging, Pharma must find a way to comply with them and participate actively in social media as it offers many opportunities. Therefore, Pharma must embrace social media. Consider these reasons, for example:

A. More and more patients are using social media as a significant source of information and an integral part of their healthcare research journey. Typically, their journey begins when they notice their symptoms. It leads to their searching social channels often before they visit a physician or simultaneously with their visit to the doctor. The second phase of online search follows the diagnosis to seek more information about their condition from credible sources and people like themselves. Patients who share their stories on social channels become a significant source of information to those actively seeking their perspectives. This information, at times, could be unbalanced and even irrelevant.

B. The information on social media may not always be from reputable and accredited sources. It may be marginally accurate at best and significantly harmful at worst.

Therefore, Pharma companies must provide balanced and credible information on social media where their patients seek meaningful information and help them get balanced, accurate information that is essential for their well-being.

### 6 Ways Pharma May Use Social Media

Dr. Kevin Campbell, an internationally recognized cardiologist, suggested six ways pharmaceutical companies can use social media in an interview with Joanna Belbey, a columnist with Forbes magazine. He said that Social Media is an ideal channel for pharmaceutical and device companies to educate, market, listen to, and connect with customers. Moreover, pharma firms can do all this while complying with the guidelines of industry regulators. Here's a summary of what he suggested:

1. *Education*: Pharmaceutical firms may offer consumers disease-specific, unbiased educational information on broad topics. Such educative information aims to help patients better understand their disease state and engage with their healthcare professionals for improved outcomes.
2. *Marketing*: Pharma can use social media to share press releases on new products, clinical trial results, community service, significant charity contributions, and their major commitment to developing new and better treatments.
3. *Connecting with Physicians*: Social media is an efficient and effective means of connecting with physician customers. It is also less expensive than sales force costs. Through social media, Pharma can engage with many physicians at once with its best and brightest scientists and researchers. It is also an excellent opportunity to ask questions and have a dialogue with each other rather than being 'detailed-to' by a sales representative, who would simply regurgitate a memorized script of a new study. The physician would have already read it in all probabilities.

4. *Connecting with Patients*: Pharma can create support groups and communities through social media. These groups allow patients with common interests and need to connect and discuss their conditions, treatments, and experiences.
5. *Clinical Trials*: Social media can become a powerful tool to enroll patients in clinical trials. Clinical trial enrollment is one of the biggest barriers to completing a clinical trial.
6. *Listening*: Social media can be a big help for Pharma in gaining insights into the unmet needs of patients through social listening. These insights can be of immense use to Pharma in developing its communication strategies with physicians and even directing the future course of its product development.

# Important Social Networks

### Physicians-only Networks

Social networks are dedicated websites or other applications which enable users to communicate with each other by posting information, comments, messages, and images. Initially, social media outlets were designed to attract individuals who use the platform for personal reasons. However, there seems to be an influx of social networks that cater to specific professions.

For doctors, physician-only social networks are becoming a valuable resource. These online communities are becoming useful resources for many professional purposes. They help physicians to communicate and collaborate in a secure environment. There is more to these physicians-only social networks than sending and receiving messages. They also provide opportunities for CME (Continuing medical education) credits and to compete against peers for various rewards. The features differ from site to site. The following table presents the top four social networks for physicians at a glance.

**Table 6.4 Top Four Social Networks for Physicians**

| Physicians' Social Networks | Event |
|---|---|
| **1. Doximity** | One of the best physician social networks. Over 80 per cent of U.S. physicians are verified Doximity members. Mobile friendly physician platform giving the ability to instantly connect with their peers in a secure, HIPAA-compliant environment. Offers the opportunity to receive CME credits for reading medical journals provided by the site. |
| **2. Sermo** | The most trusted and preferred social network for doctors with over 1.3 Million doctors across 150 countries globally. Exclusively for licensed physicians. High level of anonymity available for physicians who post on the site. Sermo allows the media to conduct polls among its physicians and interview them. |

**Table 6.4 *Contd...***

| Physicians' Social Networks | Event |
|---|---|
| **3. Quantia MD** | Launched in 2006. One of the best online resources doctors can utilize for collaboration and career development. 1 in 3 physicians visit Quantia MD to keep with new medical information and to learn from their colleagues. Has several features that reward users with CME credits or 'Q' points that can be redeemed for Amazon gift certificates. Another interesting feature is that members can compete against their peers in the *monthly general medicine puzzle, the monthly cartoon caption contest* and by solving *weekly clinical cases* and *image contests.* |
| **4. Docplexus** | Largest online community of doctors in India with over 380,000 members. Docplexus aims to provide value to the Indian medical community through innovative, technology-driven offerings. Provides audio-visual coverage for major medical events. Also offers CME courses to fulfill the educational needs of the physician community. |

Abbott is the first pharmaceutical company to launch a digital-only campaign through the physicians' network, doctors.net.uk. Consider this case.

CASE

19

# Abbott's Established Products Division Launches the First Digital-Only Campaign in Pharma!

Abbott Established Products Division (EPD) launched its low-dose hormonal replacement therapy (HRT) brand, Femoston - Conti. What is unique about this launch is that it is probably the first-ever digital-only launch in the pharmaceutical industry.

The launch campaign reached 9,000 doctors or 45 percent of the NHS population of obstetricians and gynecologists. Abbott EPD's low-dose HRT brand, Femoston-Conti (estradiol and didrogesterone tablets), engaged these doctors via the British online professional network doctors.net.uk. The campaign included interactive case studies, clinical paper summaries, and can ask the 'experts section.'

How did Abbott decide on a digital-only launch via doctors.net.uk? The decision was based on the findings of Abbott's research on the likely impact of professional networks on pharmaceutical brand promotion. Research indicated that online professional networks could provide a more effective method of engaging doctors than traditional sales and marketing channels. It is because they enable them to take a more targeted and measurable approach and deliver promotional and educational messages that readily benefit doctors' clinical practice.

Physician professional networks are increasingly becoming popular, and an estimated three million physicians use them worldwide. Statistics from doctors.net.uk show that 59 percent of them visit these channels specifically to update their medical knowledge, while 38 percent do so to get information on new products.

Abbott, already very active in the digital space, partnered with doctors.net.uk for its Femoston-Conti launch in 2013 and has won the PM Society 2013 award for the best HCP website for their Femoston-Conti low-dose created by doctors.net.uk. The award judges praised the winning entry for its excellent educational content. They said it was nice to see the industry take this step — and clearly, it was a good decision, judging by the impressive results.

The campaign was a success. Enjoy Mann, commercial marketing manager for Abbott EPD (Established Products Division), said while speaking about the campaign results to PMLiVE: *The results speak for themselves. The coverage and frequency of the campaign exceed current national call rates by sales reps. What pleases us the most is our increase in market share and continued sales growth. Doctors return to the website, indicating that they find it a valuable resource*.

(Source: Adapted from the article: Piotr WrzosiDski, Abbott EPD: The first digital marketing-only launch in Pharma, Pharma Marketing, K - Message, November 25, 2013)

### Patient Communities

Social networks focused on patients suffering from a specific disease condition have come up over the past few years. They can be vital for people who feel alone in their health struggles. They are particularly helpful to those who cannot leave home because of their disabilities. These social networks are their social life. At the same time, despite all the benefits and support they offer, patient social networks are not without challenges. Firstly, many physicians are concerned about these networks as most of the medical advice comes from other users, or unverified articles on the internet. Secondly, many patient networks do not generally moderate comments to verify their accuracy. They, however, try to make it clear where the content comes from and whether it is from medical professionals or fellow patients. The following table presents some of the more influential and important social networks that focus, and serve the needs of patients.

**Table 6.5 Some Important Social Networks for Patients**

| Patient Networks | Their Focus |
|---|---|
| **1. Patients Like Me** | Used to be primarily for some neurological conditions, but their fabulous software is available for many conditions now. |
| **2. CureTogether** | Has patient communities for a number of conditions |
| **3. SmartPatients.com** | An online community where patients and their families learn from each other. |
| **4. CancerConnect.com** | Offers comprehensive cancer information organized by disease and stage. A free social network for patients, survivors, and caregivers. |
| **5. TuDiabetes.org** | A space on the web where people with diabetes, or their loved ones can find support, help each other, and share their experiences and what they do everyday to stay healthy with this condition. |

### Online Forums

An online forum is an internet forum or a message board, or an online discussion site where people can hold conversations in the form of posted messages. Since the dawn of the internet, Online Forums have been a significant social outlet that has tremendously affected online society. As a result, they have become commonplace today.

Increasing infiltration and use of the internet and changes in the health care system have resulted in a heavier reliance on the internet for disease and health education. Every day, millions of users log on to their favorite online health communities and interact with others to get advice and discuss everything related to their symptoms, disease conditions, and treatment options. This trend has increased the expectation and reliance on peer education and support in many areas, including medicine, to precede, supplement, or even replace professionals.

Online health forums have become so popular and influential that most often, even before they call a doctor, patients go online when they are facing a health crisis or ongoing illness. As a result, more people today than ever before are turning to social media before, during, and after diagnosis. Sixty-nine percent of doctors in a survey conducted by Cello Health Insights said that many of their patients look up their conditions online before a consultation.

Online health communities, also referred to as treatment, disease state, or patient communities, represent a tremendous opportunity for pharmaceutical companies to educate patients and create goodwill among them.

By creating or partnering with online health communities, Pharma firms have an opportunity to connect with their patients and consumers. They also provide the Pharma brands an avenue to give much-needed factual information and patient-centered care and generate goodwill.

Through their partnerships with online health communities, pharmaceutical companies can help improve patients' awareness of certain diseases, which take a long time to diagnose. As a result, they can reduce the diagnostic timeline earlier in their disease journey, and this early diagnosis would ensure a better quality of life for the patients.

Understandably, a business cannot and should not be on every social network. At the same time, you cannot afford to ignore influential social networks. So, what are the right social networks for your company? To know that, you must consider what each network is good at and how you should participate in the conversation meaningfully. It is important to remember that the social media landscape changes quickly; therefore, keeping abreast of new channels is part of a good long-term social media strategy.

## Search Engine Marketing (SEM)

With a 78 percent share of the global search engine market, Google processes about a whopping 3.5 billion searches on the net every day. Also, there are over one billion websites on the world wide web today. This milestone was first reached in September 2014. Making your website stand out in this huge clutter is quite challenging. So how do you make a presence in this densely crowded cyberspace? Enter Search Engine Marketing (SEM).

Search Engine Marketing (SEM) is a form of internet marketing activity that promotes websites by increasing their visibility in search engine result pages (SERP). Search Engine Marketing covers two key tactics: Search Engine Optimization (SEO) and Search Engine Advertising (SEA).

### Search Engine Optimization

Search Engine Optimization (SEO), also known as 'organic results,' is mainly achieved by optimizing the content of websites so that they are ranked among the top of the search engine result pages. This is very important for your website because over 90 percent of all search engine traffic visits or clicks the links on the first page of search engine results. The second page receives a mere 5 percent of the traffic. Moreover, no one anymore types an URL (Uniform Resource Locator) or web address. While URL remains vital to the web's structure, web users, healthcare professionals, and patients use search engines virtually. With effective search engine marketing, you make sure you are found on the internet. Because if you are not found, you may just as well not exist.

### Search Engine Advertising (SEA)

Search Engine Advertising (SEA), also known as *Sponsored Results*, shows specific search-related ads and works on a Pay-Per-Click (PPC) basis. You can get your ads placed by a search engine in certain search results. These advertisements, however, do not appear in natural search results. Instead, they usually appear on the right-hand side of the results page in response to a corresponding search term. The fee for these ads is calculated by the number of clicks the advertisement attracts. Hence the name pay-per-click ads.

Search engine advertising is particularly useful if a website does not appear at the top of the search results because it is still new and not yet well known, or there is much competition for the search terms or keywords being used.

If Pharma marketers can't compete in the natural search results of SEO (Search Engine Optimization), they can still consider SEA (Search Engine Advertising). Here are some important advantages that SEA offers marketers:

1. Firstly, search engine advertising is based on value. You pay for clicks and not for views. Therefore, you can consider it results-driven advertising.
2. You can also consider it relevant advertising because ads are only shown when people search for your defined keywords. This differs from mass media advertising, such as television or radio, targeting everyone.
3. In addition, search engine advertising is very cost-effective. Many keywords in healthcare are very specific, so competition is generally low on those keywords. Some keywords, for example, are sold for less than ten euro cents.
4. Furthermore, you can control your costs on SEA (Search Engine Advertising). You can decide at the beginning of a campaign how much budget you want. Once you reach your budget limit, your ads will simply stop showing.
5. Search Engine Advertising (SEA) is transparently based on the auction. Therefore, the *Cost per Click* (CPC) is determined by the competition level, which is transparent to advertisers.
6. Search Engine Advertising (SEA) can provide valuable insights into your target audience's keywords, thus helping the advertisers' campaign relevancy.
7. With search engine advertising, you can also extend the reach of your campaigns. Apart from being the most visited site,

Google can also extend the reach of your campaign beyond its own website by placing your ads on relevant third-party websites. This is called Google AdSense.

8. One major advantage of search engine advertising is that people perceive text advertising as less intrusive than traditional banner advertising.
9. Above all, you can monitor your campaigns with SEA (Search Engine Advertising) in real-time and improve them continuously.

### Which CPC to Measure?

Which is the ideal metric to measure when using SEA (Search Engine Advertising)? Cost-per-Click (CPC) or Cost-per-Conversion (CPC)?

Many marketers use cost-per-click as a metric to measure a campaign's success. However, the number of clicks does not tell about the quality of the visitors and visits. Instead, consider the goals of your campaign. What do you want your visitors to do once they are on your web page? How do visitors convert? Do they download a factsheet, a pdf of an article, register for an event, or subscribe to your newsletter and so forth. You can track these activities, calculate the conversion cost, and compare those across different channels.

### Search Engine Marketing (SEM) versus Display Advertising

Search engine marketing (SEM) and Display advertising (banner advertising) are distinctly different. SEM is a pull tactic. It pulls people in. Display advertising is a push tactic. It pushes ads toward people. Display advertising is used more frequently in the early stages of the customer journey when they are not actively looking for a solution to a problem. On the other hand, search is more frequently used in the latter stages of the customer journey, such as closer to the treatment, product decision, and drug adherence. Google Analytics can help you monitor your brand's performance across channels in real-time.

### SEM and the Customer Journey

Understanding how patients and healthcare professionals search throughout the customer journey is essential. Studying your customers' search behavior will give you actionable insights and enable you to align the content of your website and your search engine marketing strategy. In addition, it will help you make your content more relevant for healthcare professionals and patients, and place it more prominently on the result page of search engines.

### Search Engine Optimization (SEO)

You need to optimize your website and content if you want to be found on the world wide web. And it is very difficult to be found on the web with over a billion websites. It is not enough if you are found on the web. You have to be listed on the first page of the search engine results page (SERP). Because if you are not on the first page of search engine results, you will likely get a mere 5 percent of the traffic. Therefore, optimizing your website and content is no longer a matter of choice. It is rather mandatory. Search engine optimization (SEO) can optimize your website and other digital channels for search engines. Search engines are the starting point for everything for most web users.

### How Search Engines Work

A search engine works basically through a web crawler, also called a web spider. A web crawler is an automated software application that systematically browses (or crawls) the world wide web. Web crawlers copy all the pages they visit and store them in a web index. Because of this web index, a search engine can generate result pages in a split second. When we perform a search, the search engine flips through this index, which is like a massive book, locates all the relevant pages, and selects the pages it thinks are the best to show first. The algorithm defines the selection logic.

Search engines create their result pages based on a series of algorithms. These algorithms determine whether your page gets a high ranking or not. Therefore, you need to monitor the positioning of your web pages and your competitors continuously as search engines update their algorithms, which can impact the ranking of your web pages. With most traffic (over sixty-one percent) coming via search engines, a position drop can significantly impact your traffic and, hence, your business.

### Choosing Keywords

Choosing the right keywords is crucial for success in search engine marketing. The keywords you use in your content determine your rankings on search engine result pages (SERPs). To help you select the right keywords, Google has developed a Keyword Planner tool. It generates a list of relevant keywords and monthly search volumes for a therapeutic area or product. The tool helps you select keywords for search engine optimization(SEO) and search engine advertising (SEA).

### SEO Success Factors

Surprisingly, only 42 percent of pharmaceutical companies actively optimize their websites to improve their search engine visibility. However, search engines are the starting point for most web users to search for everything. Here are ten critical success factors for optimizing your search engine and improving your position on its result pages:

1. *Value of Keywords*. SEO is free. You don't have to pay to be shown on search engine result pages (SERPs). The keywords you choose and use in your content determine the position of your website and other channels in search engine result pages. Selection of the right and most relevant keywords that match your customer's search does take time and effort. Your keywords are very valuable.
2. *Content is king*. Search engine optimization (SEO) is content-driven. You optimize your content for search engines by incorporating the most relevant keywords. Besides, you also need to update your content at regular intervals continuously. The more relevant and recent your content is, the greater the chances of higher rankings in search engine result pages. The higher your content appears in the result lists, the greater the probability of visiting your content.

3. *Technology is Important*. In addition to relevant keywords, technology is crucial for optimizing your search engines. You must ensure that your website is easily accessible to search engines. Optimizing your site structure to facilitate both surfing and crawling is essential. Search engines only interpret the text, and content tagging (text descriptions) enables them to interpret and index those visual elements. It is, therefore, important to tag all the visual elements on your sites, such as images, videos, PDFs, and others. Also, optimize your website for mobile. Mobile-optimized websites get better positioning in search engines.
4. *Credibility is crucial*. Credibility ensures better rankings on search engine result pages. One simple way to improve the reliability of your website is to have other websites link to your site. When other important websites link to your site, that enhances your site's credibility leading to an improved ranking in search engine results. Remember to enable social sharing because these additional links help you improve your page rankings and traffic to your site.
5. *Optimize your website*. To optimize your website, you first need to create quality content. Sites with interesting and engaging content will have better search engine traffic. Easy-to-read text that is frequently updated and appropriately tagged content is highly engaging.
6. *Optimize linking*. Link all your other websites and also optimize them. The more links are, the more clicks.
7. *Combine channels*. Combining channels facilitates multichannel engagement and can create a seamless experience.
8. *Make your content shareable*. People share content they find useful, trendy, engaging, credible, and trustworthy. Also, make it easy to share by creating social sharing links.
9. *Promote your website*. The more traffic the websites get, the better position they will get in the search results.
10. *Monitor*. Monitor your search engine rankings continuously. When search engine algorithms change or competitors become more active, your position on the search engine results may fall.

## SEO in Pharma Industry

Pharmaceutical companies, being highly regulated, face some challenges in implementing their search engine optimization strategies. While these are not insurmountable, pharmaceutical marketers should consider them when planning their SEO strategies.

- *Copy approvals* for all types of content in the pharmaceutical industry take time as they involve multiple functions, such as legal, medical, and regulatory. Keeping content fresh and up-to-date can, therefore, be challenging. It is essential to start the search engine optimization simultaneously as soon as content development starts.
- *Right Keywords*. You may not include the most appropriate and relevant keywords, such as your product names, as organic search on your website as they will not be allowed.
- *Password challenges*. Most websites for healthcare professionals are password-protected. However, search engines cannot index password-protected content. Search engines, therefore, index only the landing pages of HCP sites, and due to this limited access to your content, these sites are unlikely to show up high in search results. What is the alternative? How can you improve your page ranking? Include as much descriptive content as possible on the landing page of every password-protected site. You can also add a section for patients that can be publicly accessed from your landing page.

Analyzing the traffic to a website and other channels is crucial to achieve success in SEO (Search Engine Optimization) and Search Engine Advertising (SEA). Google Analytics can help you analyze conversions, optimize keyword bid prices, and look at bounce rates and the cost-effectiveness of your campaign.

Finally, continuous optimization of your campaign as it is running holds the key to success. You can add relevant content to your website, identify new keywords, and optimize budgets accordingly.

# Artificial Intelligence (AI), and Machine Learning (ML) in Pharma Marketing

Although research into artificial Intelligence (AI) started in the 1950s, the industry has only recently used AI terminology. However, according to Veeva, one of the most successful and innovative cloud software companies, AI technology will soon transform pharmaceutical sales.

### What is Artificial Intelligence?

Artificial Intelligence (AI) is the development of intelligent computers that can perform several human-like tasks. The technology should develop reasoning, problem-solving, perception, planning, and learning from past experiences. Finally, artificial intelligence (AI) should think and behave rationally like humans. AI has progressed significantly and is now integrated into our daily lives, which we can experience with Apple's Siri and Amazon's Alexa. Even though these systems perform specific tasks well, they require direction and are considered weak. Artificial Intelligence (AI) aims to develop robust systems to make decisions and intelligently handle various situations.

### Machine Learning (ML)

Machine learning (ML), another buzzword in the area, is an artificial intelligence (AI) branch. There is one difference, however. In contrast to AI, which imitates human capabilities, machine learning (ML) trains computers using vast data to recognize patterns and make decisions without direction. Machine learning (ML) algorithms have been around for some time, but a recent development can now apply machine learning to big data. Netflix offers an excellent example of how to use machine learning. It uses technology to tell you which film to watch or a series to binge on next.

### Many Benefits of AI and ML

While artificial Intelligence, with all its capabilities, remains a work in progress like any other evolving technology, many industries are already reaping its benefits. For example, banks use AI to cut costs, increase efficiency, and increase security. In addition, improved voice and chatbots offer an alternative to human customer service and potentially save the industry up to $450 billion by 2030. Besides, you can train the pattern recognition capabilities of machine learning to spot fraudulent activity in real time, making it an asset in the fight against cybercrime. Also, AI and ML have made self-driving cars a possibility, with Tesla leading the way with its autopilot hardware. Similarly, Japanese shipping companies are developing self-navigating vessels using AI to calculate the safest and most fuel-efficient route to their destinations.

**Artificial Intelligence (AI) and the Healthcare Industry**

Artificial Intelligence (AI) has also been helping the healthcare industry in many ways. For example, in the oncology segment, in 2018, cancer caused 9.6 million deaths worldwide. AI technology now helps oncologists diagnose cancer early, making safer treatment possible, allowing fast and accurate biopsy analysis, and reducing error and the need for repeat tests. Also, an algorithm is being developed to identify lower, tumor-shrinking doses of drugs that provide effectiveness with reduced toxicity.

**Artificial Intelligence and Pharma Industry**

Artificial Intelligence (AI) has already proved itself as a useful tool to support the research and development activity carried out by pharmaceutical companies. However, with increasing investments in AI, Pharma companies are exploring ways to exploit its potential in marketing. In January 2019, Paul Shawah, senior vice president of Veeva Commercial Cloud, opined that AI would significantly impact the life sciences industry in 2019. He said: *We will move towards an AI-driven workforce as artificial Intelligence becomes a standard capability across commercial applications. AI will not only continue to improve sales force effectiveness with suggestions on their next best activities... It will also automate references-linking in content management for marketing operations.*

According to Accenture, one of the leading global strategy consulting firms, AI could help workers to use their time more efficiently, increasing labor productivity by up to 40 percent — just the sort of boost that sales reps need.

The impact of AI on pharmaceuticals can start before the sales rep meets a customer. Machine learning (ML) can separate customers into highly specific segments allowing sales teams to personalize their activity to a higher degree. You can, for example, segment doctors based on factors such as interest in new drugs, location, availability, therapy areas, and specialization.

Artificial Intelligence (AI) and Machine Learning (ML) are rapidly becoming essential technologies in the pharmaceutical industry, particularly in marketing. These technologies offer a wide range of benefits that can help pharmaceutical companies improve the effectiveness and efficiency of their marketing efforts. Consider these benefits, for example:

- Pharmaceutical companies collect vast customer data, including demographic information, medical history, and purchase history. Artificial Intelligence and Machine Learning can process and analyze this data to identify patterns and trends, which can be used to target marketing efforts more effectively. For example, machine learning algorithms can predict which physicians are most likely to be interested in a particular drug and target those customers with personalized marketing campaigns.
- Another benefit of AI and ML in pharma marketing is their ability to automate repetitive tasks. Many marketing tasks, such as data entry and analysis, are time-consuming and labor-intensive. AI and ML technologies can automate these tasks, freeing marketing teams to focus on more strategic activities. For example, natural language processing (NLP) algorithms can automatically extract information from scientific papers, which can be used to create content for marketing campaigns.
- AI and ML can also play a major role in personalizing the customer experience. Pharmaceutical companies can use these technologies to create personalized content and customer recommendations. For example, AI algorithms can

analyze a customer's medical history and current medications to recommend other drugs that may be beneficial to them. Personalized marketing campaigns can also be created using AI algorithms to target specific groups of customers based on their demographics, behavior, or medical history.

- AI and ML can also be used to improve the efficiency of marketing campaigns. Machine learning algorithms can optimize marketing campaigns' targeting, timing, and messaging to maximize the return on investment. For example, AI algorithms can predict the most effective time to send an email or message to a customer based on their behavior and engagement history.

AI and ML offer pharmaceutical marketers some of the more important benefits. As a result, pharmaceutical companies can improve the effectiveness of their marketing efforts and better serve their customers with these technologies.

## Agile Marketing in Pharma

Agile marketing is a flexible and adaptive marketing approach characterized by continuous experimentation, learning, and iteration. It is based on the principles of agile software development and emphasizes cross-functional collaboration, rapid prototyping, and customer-focused decision-making. Agile marketing teams work in short sprints to test and refine marketing strategies and tactics, and they continuously gather customer feedback to inform future iterations. As a result, agile marketing aims to quickly and efficiently deliver value to customers while maximizing the impact and ROI of marketing efforts.

Agile marketing focuses on the customer, validated learning, short adaptive iterations, and cross-functional collaborations. Thus, it differs significantly from traditional marketing. Jim Ewel, one of the leading voices of Agile and a co-author of the book *Agile Marketing Manifesto,* highlights the important differences between traditional and Agile marketing. Here are the key differences:

1. Traditional marketing has an inward focus. It focuses on the producer and the product. The four Ps (Product, Price, Place, and Promotion) are inward-looking levers under the producer's control. In contrast, Agile marketing focuses on the customer and the customer value chain. For example, Robert F. Lauterborn's Four-C model, where customer experience replaces the product, cost replaces price, convenience replaces place, and communication replaces promotion, is more appropriate for Agile marketing.
2. Traditional marketing focuses on the producer's sales cycle, and Agile marketing on the customer's buying process.
3. Traditional marketing focuses on the Spin, controlling the message, and owning the brand. Agile marketing focuses on transparency and engagement and recognizes that, ultimately, customers own both the message and the brand, and only through honest interaction can a company influence customers.

4. Traditional marketing relies on measurements like Total Addressable Market (TAM), market share, unaided recall of a brand or ad, Gross Rating Points (GRP), and Target Rating Points (TRPs). In contrast, Agile marketing proceeds via a series of *experiments* forming a hypothesis of what might work and testing that hypothesis. Agile marketing chooses measurements appropriate to the hypotheses at hand and measurements that align with the organization's business and sales goals. Traditional marketing is measured in terms of success and failures, whereas Agile marketing is measured in terms of the pace of learning.
5. Traditional marketing believes that markets and customers are like a lump of clay you can mold into their idea of what the customer and the market should be with the marketing tools (four Ps, advertising, and public relations). In comparison, Agile marketing treats markets and customers as a complex system of individuals trying to learn what works for them and is not surprised when they don't react as expected.
6. The traditional marketing philosophy is to find a big idea, get it right, and launch it with a big splash. It relies on a series of campaigns, yearly, quarterly, and big launches. In contrast, Agile marketing focuses on a series of short Sprints, emphasizing getting a good enough idea, testing it, and perfecting it over time.
7. Traditional marketing is often organized around internal functions such as Public Relations, Advertising, Product Management, Search Engine Optimization (SEO), and Search Engine Marketing (SEM). Conversely, Agile marketing tends to be organized around the customer value cycle (need recognition, information search, evaluation of alternatives, purchase decision, and post-purchase behaviors). In Agile marketing, different functions often sit in the same space and collaborate to meet customer needs.

### Agile Marketing: What It Means

Agile marketing involves short-term planning and speedy execution cycles—typically a few weeks, allowing marketing teams more visibility and flexibility in prioritizing tasks and meeting objectives. In addition, agile marketing uses the bite-sized decision-making approach, allowing a course correction to be on target due to its iterative process.

Agile marketing also means using data and analytics to guide opportunities and solutions to problems, continuously developing tests, quickly evaluating the results, and rapidly iterating.

Furthermore, Agile marketing enables quick decision-making and seamless collaboration by breaking down organizational silos with its cross-functional team approach.

### Breaking Silos

Pratap Khedkar, chief executive officer of ZS Associates, a leading global management consulting and professional services firm, summarizes the Agile approach in three steps: *Anticipate, Decide, and Act*. And to capitalize on those actions faster, some pioneering pharma companies are developing ways to break down the communication silos to foster an Agile approach.

How can pharma implement agile marketing practices? Here are eight steps to implementing agile marketing in pharma:

1. *Assess current process and culture*: Assess the current marketing processes and culture to identify where agile marketing could be implemented.
2. *Cross-functional collaboration*: Encourage collaboration and cross-functional teamwork between departments, including marketing, sales, MLR (Medical, Legal and Regulatory), and product development.
3. *Agile methodology adoption*: Adopt an agile methodology, such as Scrum or Kanban, to help guide the marketing process and ensure that work is prioritized and delivered efficiently.
4. *Customer-focused approach*: Focus on customers and prioritize their needs in all marketing activities. This can be achieved through regular feedback loops, customer research, and testing.

5. *Flexible planning and execution*: Plan and execute marketing activities iteratively and flexibly, allowing for changes as needed.
6. *Continuous improvement*: Continuously review and assess the effectiveness of marketing efforts, and make changes as needed to improve performance and meet customer needs.
7. *Empower teams*: Empower teams with the autonomy and resources to make decisions and take ownership of their work, leading to increased creativity and innovation.
8. *Adaptation to change*: Be prepared to adapt quickly to changes in the market, customer needs, and regulations and continuously refine the agile marketing process to ensure its effectiveness.

Several pharmaceutical companies are practicing agile marketing, and here are some examples of how they are doing it:

1. Pfizer has adopted Agile Marketing to rapidly test and iterate their marketing strategies, allowing them to quickly respond to changing market conditions and customer needs.
2. Johnson & Johnson uses Agile Marketing to develop more flexible and adaptive marketing campaigns, focusing on customer-centered innovation and continuous improvement.
3. Novartis has implemented Agile Marketing to improve collaboration and speed in their marketing operations, focusing on customer-centered product development and faster time-to-market.
4. Sanofi has adopted Agile Marketing to understand customer needs and preferences better and, rapidly test and iterate its marketing strategies to meet these needs.
5. Roche has embraced Agile Marketing to enhance collaboration and flexibility in their marketing efforts, focusing on customer-centered product development and rapid prototyping.

## Design Thinking in Pharma

Design thinking is important in pharma marketing because it helps companies understand and meet the needs of their customers in a creative and empathetic way. In addition, design thinking is of significant help for pharmaceutical companies in creating marketing strategies that are more effective and resonate with their target audience because of its human-centered approach.

The seven important steps that pharma should consider while implementing design thinking are:

1. *Empathy*: Understanding the customer's needs and pain points through research, observation, and customer feedback.
2. *Ideation*: Encourage creative thinking and ideation through workshops, brainstorming sessions, and cross-functional collaboration.
3. *Rapid Prototyping*: Create prototypes and test ideas quickly to validate assumptions and gather feedback.
4. *Testing and Iteration*: Create prototypes and test ideas quickly to validate assumptions and gather feedback.
5. *Customer-centered Focus*: Keep the customer at the center of all marketing efforts, making decisions based on their needs and feedback.
6. *Collaboration*: Foster collaboration and cross-functional teamwork between marketing, product development, and other departments.
7. *Continuous Improvement*: Continuously review and improve the marketing process to ensure it is effective and efficient and to stay ahead of the competition.

### Patient Focus is the Key

While implementing design thinking, pharmaceutical companies should focus on their patients because putting patients at the center of their focus to gain a deeper understanding of their needs and wants will significantly help the design and development of products and services.

## Design Thinking in Pharma

Pharmaceutical companies are using design thinking in several ways, including:

1. **Drug Delivery:** Design thinking is used to develop innovative drug delivery systems, such as inhalers and patches, that are easier to use for patients.
2. **Clinical Trial Design**: Design thinking helps pharma companies better understand patient needs and preferences, leading to more patient-centered trials.
3. **Packaging Design**: Design thinking is used to create more intuitive and user-friendly packaging, improving patient adherence to their medication regimens.
4. **Service Design**: Pharma companies are using design thinking to design patient-centered services, such as telemedicine and remote monitoring, to improve access to care and patient outcomes.

## How Pharma is Using Design Thinking

Here are some specific examples of how pharmaceutical companies are using design thinking:

1. Pfizer used design thinking to develop a more patient-friendly inhaler for their COPD medication, with features like a simple and intuitive design, clear instructions, and a small and portable size.
2. Johnson & Johnson used design thinking to create an insulin delivery system that is more user-friendly and less intimidating for patients. The design features a discreet and compact form factor, a simple and intuitive user interface, and real-time insulin use tracking and analysis.
3. Novartis used design thinking to develop a mobile app that helps patients manage their multiple sclerosis (MS) medication regimens. The app features a simple and intuitive design, personalized reminders, and real-time tracking of symptoms and treatment.

4. Sanofi used design thinking to create a remote monitoring system for their diabetes patients. The system features a wearable device that tracks blood sugar levels, a mobile app that provides real-time information and insights, and virtual consultations with healthcare providers.

## Content Marketing

The internet is the starting point for most people researching health information. Digital Pharma reported that over 12 months, 72 percent of internet users mentioned in a survey that they looked online for health information. Pharmaceutical companies can use this opportunity to create informative and engaging content on health information related to their therapeutic areas and brands.

Who and what rules the internet? Content, of course! Content is king when it comes to marketing online. Content is the fuel that drives your digital marketing activities, such as Search Engine Optimization (SEO). Content is what gets you noticed on social media. Content gives you something of value to offer your customers in emails and paid search ads.

Creating content that is not promotional but educates and inspires instead, gets you noticed. Content that offers valuable and relevant information ensures that your customers stay tuned in. The key to good content is that it avoids being overtly promotional, and is useful or valuable to the customer. Because if the content is not relevant to the consumer, whether it is the patient or HCP, they go elsewhere. They want to understand the condition and the treatment options available to them. Therefore, you need to give the information they need to decide on a language they fully understand, through the medium they want, and when they want it. Quality of content is crucial to engaging them on their terms. The excellent quality content aims to create a two-way engagement rather than delivering a one-way sales pitch. Once you generate good quality content, you need to market it effectively. How do you market your content? Content marketing, as defined by the *WhatIs* the website, is: *Content marketing is the publication of material designed to promote a brand (or service), usually through a more oblique and subtle approach than that of traditional push advertising. The essence of good content marketing is that it offers something the viewer wants, such as information or entertainment*. Content marketing can take many forms, including Youtube videos, blog posts, and articles. Sometimes, it shouldn't seem like marketing; it should only be identified because the advertiser is the content provider.

## Modular Content

In the pharmaceutical industry, marketing and sales teams can build on the existing content and adapt it to suit the new markets while retaining the essential elements consistently. With a modular content approach, pharma marketers can repurpose their content across markets and channels, ensuring the same quality.

What is modular content? Modular content is creating pre-designed blocks of content modules to create marketing assets. These modules cover all the elements required to provide a detailed picture of the content, including product benefits, claims, references, graphics, and logos. In addition, each module has specific business rules on how to use it, resulting in the content you can deliver as a standalone or in combination with other content modules.

Creating modular content helps pharma marketers in developing content quickly and reduces inefficiency. In addition, modular content allows scientific content management facilitating content reuse to create effective omnichannel content quickly. The important benefits of modular content are:

- The sales and marketing teams can engage with customers more effectively by having marketing assets ready.
- Pharma marketers can develop MLR (Medical, Legal, and Regulatory)-approved content modules without MLR constantly reviewing marketing assets. As a result, they can focus more on new features and experiences, repurposing the content more effectively.
- Creating content structure in advance allows users to reinforce customer perception about the brands with consistent content across multiple channels, creating an omnichannel experience. Furthermore, modular content allows you to create effective omnichannel strategies since you can assemble the content modules quickly and in different configurations to build personalized customer journeys, allowing you to tell your product story across channels.

### Creating Modular Content: Important Steps

How do you develop modular content that is consistent across channels? Consider the following steps.

1. Create a generic content template with core brand assets, so your teams can generate customized content faster per promotional strategies.
2. Make the content creation process consistent across all teams through documentation.
3. Involve your MLR (Medical, Legal, and Regulatory) team early in the modular content creation process, so everyone is onboard and understands the nuances of promotional strategies and possibilities.
4. It is important to assess the technology requirements for integrating modular content. For example, you need a digital asset management solution where content is stored, shared, and managed. Also, a content authoring tool for modular content can help create and manage content faster.

### Optimize Your Content

It is not enough to create great content. You must optimize it because a strong online content marketing strategy is also the key to SEO (Search Engine Optimization). Google's algorithm keeps evolving to promote web pages with high-quality content better. But, as Karim Binon, digital marketing manager of Oceasoft, a leading provider of high-quality temperature monitoring solutions based in Montpellier, France, rightly observed: *Posting great content on a technically not optimized website is like riding a Harley Davidson equipped with a scooter's engine. It won't go very long, and not very fast*.

Pharmaceutical marketers must provide engaging content that meets the needs, wants, aspirations, and goals of all their key stakeholders, namely, patients, HCPs, providers, and payers. Meeting the needs and interests of the target audience is crucial for success. Pharma should create different content for each customer segment in a way that speaks directly to their particular issues. A one-size-fits-all approach is outdated now. The key questions to ask and answer are:

- Who is your target audience?
- Are they patients or general practitioners? Specialist physicians?
- What information do they want?
- What issues do they have, and how can you solve them?

Remember to be engaging; content should be interactive, interesting, informative, and shareable. Your content should not seem like a marketing strategy or be promotional. However engaging, content cannot promote itself unless at least you share it on social media.

A successful and sustainable content strategy for a pharmaceutical company hinges on developing a trusted relationship using the content. Remember, trust is an emotion based on credibility. It can be built only on aligned values. Therefore, pharmaceutical marketers must be very objective in developing their content and focus more on building credibility than grabbing attention. Of course, you need to grab attention and sustain it, but credibility is more important in building a trusting relationship. In healthcare, it is even more critical. You can draw attention and build trust only when you have a strong point of view based on facts in your content. Otherwise, it would be a series of facts that anyone can find with a Google search. Your content must resonate with all your key customers— patients, physicians, and payers. All of them should feel that the pharmaceutical company understands their values and beliefs, and that it is not just educating them on a prescriptive solution.

Ramya Sriram outlined the seven essential steps in the content marketing strategy titled *Content Marketing for Biotech & Pharma: The Ultimate Guide*, written by Christopher Oseh, an experienced science content writer. Here is a gist of the seven-step process of content marketing in the life sciences industry:

1. *Content Goals*: Set specific goals you want to achieve with your content, set milestones for your goals and track the progress.

2. *Define Your Audience*: Outline specific characteristics such as pain points, challenges, problems, and needs that match your audience to determine the type of content to create so that your audience engages optimally with your content.
3. *Content Research and Plan*: Use appropriate research tools to discover the topics of your audiences' interest and schedule a calendar for publishing the selected topics.
4. *Intentional Content Creation*: Create search engine-optimized content that communicates with your audience at every stage in the customer's journey.
5. *Content Distribution*: Decide on the appropriate channels where your ideal audience can easily see and engage with it.
6. *Content Performance Analysis*: Analyze all published content periodically to identify top-performing content and focus only on them. Measure common indices like the number of unique visitors, likes, shares, comments, and leads.
7. Types of Content: As there are many types of content you can create to engage your audience effectively, matching the content type with the audience's needs and preferences are important. Here are some types of content that your audience would be interested in:
   - *Case Studies* are a type of content that involves interviewing customers who have used your products and services. It involves asking the right questions, using multimedia content, and creating a resource page to engage the audience effectively.
   - *White Papers* are compelling special reports written to solve problems faced by a target audience, usually written by subject matter experts.
   - *Press Releases and Blog Posts* are a great way to communicate any event, product launch, company news, or industry-related discovery to the public.
   - Social Media Posts are more suited for short, catchy content with well-designed images and short videos to improve interaction with your audience.

## A Sage Advice

Robert Rose, the chief strategist at Content Marketing Institute and co-author with Joe Pulizzi of the book, Managing Content Marketing, advised that:

*Content marketers should start thinking of themselves as media companies— with a distinct point of view. Their goal is to develop an audience for their content*. However, they must balance that with how, when, and where they will show their distinct and unique point of view.

# Closed Loop Marketing

## Evolution of Closed Loop Marketing

First, Pharma started extending customer relationship management (CRM) with paper-based promotional materials to promote their products. Later, it started with the advent of digital technologies digitizing paper materials and placing them on mobile devices—first laptops, then iPads. Moving from paper to digital or CRM Plus meant even more efficiency and, in turn, achieved cost savings on material localization, and distribution.

However, CRM Plus used the same approach that had been around for a long time, the only difference being that materials were more digital than paper. In addition, Pharma was still trying to push messages to customers, hoping that some of them would stick. So while there were some cost savings, fundamentally, nothing had changed. It was business as usual— pushing messages. The truly revolutionary step came next— *Closed-Loop Marketing (CLM)*.

What is different with CLM? Instead of firing out messages to the healthcare professionals and hoping that some would stick and work, CLM comes with insight to understand people's reactions.

However, the standard CLM begins by engaging with customers and recording their reactions, enabling companies to adapt what they are doing to reinforce the customers' responses positively. It leads to a new cycle of engaging, understanding, adapting, engaging, and so on.

Digital and multichannel marketing are often tied to closed-loop marketing (CLM) because CLM can integrate different digital channels, such as websites and social media, as part of the detailing efforts of a firm.

Pharmaceutical companies identify and refine the marketing promotional materials their physicians find exciting and resonating. Then, they seek to adapt their promotion efforts and direct them back to the physician, closing the loop. Consider, for example, the following steps a Pharma company may take in implementing closed-loop marketing.

A pharmaceutical sales representative (PSR) is provided with an iPad or tablet containing the detailing software. This software enables the PSR to present an engaging, detail-talk about his product through a multimedia presentation that may include PDFs of clinical article reprints, a slide deck, and a video featuring a key opinion leader or mode of action, the drug or interactive graphs, among others. While the physician is engaged in the detail talk and viewing these materials, a record is made of which ones are used and what length of time was spent on each page, slide, graph and video.

The recorded information is sent immediately to a central system at the firm's head office via the internet. The feedback data from the tablet is analyzed with the customer relationship management (CRM) data that the company has (companies that implement CLM strategy will most certainly have CRM in place). The insights derived from this integrated analysis are used to prepare content that meets that particular physician's needs, which is then communicated to the PSR and brand manager. PSR delivers the message via the preferred channel of the physician and at the time and place the physician wants. This process ensures the delivery of the right message that the physician can resonate with at the right time, on the right channel, and at the right place.

Thus, CLM (Closed Loop Marketing) can be a powerful way to improve customer-centricity by incorporating real-life customer insights into marketing strategy and messaging in real-time. In the same way that consumer retail websites update customer recommendations based on past activity, Pharma can customize Healthcare Professional (HCP)-marketing strategies and messages based on individual preferences learned from past customer interactions. For example, pharmaceutical companies implement CLM in the sales representative channel via tablet detailing, capturing insights from the representative's interactions with HCPs, such as the content that the HCP found most engaging electronically. Pharma marketers then customize future messaging for HCPs based on those insights.

At the same time, Pharma has been exploring the use of channels beyond the traditional field force, such as tele-detailing, video-detailing, physician communities, websites, email, and others. Pharmaceutical companies hope to increase and improve physician engagement by offering multiple interaction options through smaller, single-channel pilots and some coordinated MCM (Multichannel Marketing) programs.

With these insights derived from gathering and analyzing customer interactions from real life, not from marketing research, the brand manager may develop branding strategies nationally or for specific geographical regions based on physician preferences. Combined with meaningful segmentation, these insights allow accurate customization of content quickly achieved in digital channels and allow pharma marketers to send the right message and the most relevant content to customers via their preferred channels.

When you practice closed-loop marketing, you are engaging the physicians optimally because there is continuity in conversations. Every interaction becomes meaningful as it takes the physicians' needs, concerns, and wants into account and addresses them promptly.

**CLM in Pharma: What it Takes**

Closing the loop in CLM is about clearly understanding the efficiency of your marketing efforts. The first step to achieving this is reaching out to your customers for feedback and evaluating its impact.

The growing popularity of tablet detailing reinforced and increased the implementation rate of closed-loop marketing in Pharma as they enable users to bring in feedback, gain insights, analyze data, and adjust the next best steps leading to effective engagement. Closed-loop marketing helps meet the most important needs of HCPs. Consider these, for example:

- HCPs need immediate and around-the-clock access to the necessary information to maintain an effective practice and provide quality patient support. Digital information solutions can meet these needs by improving regular practice and continuous learning.

- Another important service that HCPs need is accurate and unbiased information for the patient. The multichannel closed-loop enables real-time access to fully customized and relevant medical information.
- Pharmaceutical and life science companies need personalized information delivery to improve their HCP engagement. They often combine traditional sales models, such as calls and face-to-face meetings, integrating them with multiple channels, including web-based sampling and social media.

### Technology Behind CLM

Technology advancements drive the closed-loop marketing processes to meet the HCPs' needs so they can serve patients better. Here is an illustrative list of the technical elements used in the CLM process as suggested by The Avenga Team in their blog article, *CLM in Pharma: The Concept*, published in avenga.com on December 27, 2020:

- Software applications for end-users and customers, such as contact centers, social media applications, tablets, mobile device specifications, etc.
- Data flow management includes activity tracking, HCP and customer targeting, and channel content distribution.
- Data repositories for HCP activity and customer data management, and as part of an enterprise data warehouse.
- Integration with other software systems, such as SFA (Sales Force Automation), ERP (Enterprise Resource Planning), CRM (Customer Relationship Management), HCP Aggregate Spend, etc is important.

### Implementing CLM

In addition to improving the effectiveness of customer engagement, the CLM system also provides complete transparency of pharma marketing expenditures and sales indicators. However, CLM cannot by itself improve engagement and customer experience. Here are some essential points to ponder and act upon while implementing CLM.

- Data and statistics obtained through CLM are not actionable by themselves. Companies need to process the retrieved data to make it actionable.
- Transferring data to a mobile device is not much of use without a well-established customer engagement experience and interaction processes. Therefore, designing a customer experience process that can drive engagement for implementing CLM is important.
- It is important to train your team with regular follow-ups to integrate CLM into your marketing effectively, and overcome any possible resistance by HCPs and your internal customers.
- Highly relevant content is crucial for the successful implementation of CLM and for engaging HCPs meaningfully because the success of CLM strategy depends on well-developed targeted content, such as eloquent case studies, relevant data, testimonials, rich graphics, interactive videos, and smooth navigation.

Closed-loop marketing thus can help pharmaceutical firms prepare communication strategies that their customers find compelling. With CLM, we can give customers what they want, how they want it, and when they want it.

## Data-Driven Marketing in Pharma Marketing

Data-driven marketing in the pharmaceutical industry uses data and analytics to inform and guide marketing strategies and decisions. This can include using data to identify target audiences, developing personalized messaging and content, and measuring the effectiveness of marketing campaigns.

The use of data in the pharmaceutical industry has been increasing in recent years as companies seek to gain a deeper understanding of their customers and improve their marketing efforts' effectiveness. This can include using data from various sources, such as social media, customer surveys, directly from patients through their wearables, and clinical trials.

One key aspect of data-driven marketing in the pharmaceutical industry is using real-world data (RWD) to inform the development and marketing of drugs. RWD refers to data collected from patients outside clinical trials, such as electronic health records (EHR), claims data, and patient-generated data from wearables. This data can provide valuable insights into the real-world effectiveness of drugs, including patient outcomes and treatment patterns.

Another aspect of data-driven marketing in the pharmaceutical industry is using digital analytics to track and measure the effectiveness of digital marketing campaigns. This can include web analytics, social media analytics, and search engine optimization (SEO) to track website traffic, engagement, and conversion rates.

The use of data-driven marketing in the pharmaceutical industry can improve the effectiveness of marketing efforts and enhance the patient experience. However, it also raises important ethical considerations, such as protecting patient privacy and the potential for biased decision-making.

## One-to-One (Personalized) Marketing

One-to-One marketing, also known as personalized or individualized marketing, is a strategy that targets specific individuals or groups of individuals with tailored messages and offers. This approach can be highly effective in reaching and engaging patients, healthcare providers, and other stakeholders in the pharmaceutical industry.

One of the main benefits of one-to-one marketing in the pharmaceutical industry is the ability to target specific patient populations. For example, a pharmaceutical company could develop targeted marketing campaigns for patients with a specific condition or demographic, such as older adults or individuals with a certain genetic profile. This allows the company to deliver messages and offers that are more relevant and more likely effective for those patients.

Another benefit of one-to-one marketing in the pharmaceutical industry is tracking and analyzing customer data. By collecting patient behavior and preferences data, companies can gain insight into which marketing strategies are working and which are not. This information can then be used to make data-driven decisions and optimize marketing efforts.

One-to-one marketing in the pharmaceutical industry also allows companies to build stronger relationships with patients and healthcare providers. By tailoring messages and offers to specific individuals, companies can create a sense of personalization and engagement that can increase trust and loyalty.

However, one-to-one marketing in the pharmaceutical industry also presents some challenges. Consider these challenges, for example:

- One of the main challenges is ensuring compliance with regulatory requirements. Therefore, pharmaceutical companies must ensure that all marketing efforts comply with laws and regulations regarding promoting drugs and medical devices.
- Another challenge is the cost. One-to-one marketing can be expensive, as it requires the collection and analysis of data,

the development of targeted messages and offers, and the use of specialized technology.

## How Big Pharma Does One-to-one Marketing

Several big pharma companies have successfully implemented one-to-one marketing strategies. Here are a few examples:

1. AstraZeneca is a global pharma major that has used one-to-one marketing to reach and engage patients with specific conditions. For example, they have developed targeted campaigns for patients with asthma and COPD (Chronic Obstructive Pulmonary Disease), using personalized messages and digital tools to deliver relevant information and resources to those patients.
2. Merck is another leading global pharma company that has used one-to-one marketing to reach specific patient populations. For example, they have developed targeted campaigns for patients with diabetes, using personalized messaging and digital tools to deliver highly relevant messages.
3. Novartis, a global pharma major, has used one-to-one marketing to build stronger relationships with healthcare providers. It uses targeted messaging and offers to reach oncologists and primary care physicians and provide them with relevant information and resources to support their patients.
4. Pfizer, a leading pharmaceutical company, has also implemented a one-to-one marketing strategy with targeted social media and email campaigns to reach specific patient populations, such as cancer patients, and provide them with relevant information and resources to support their treatment.
5. Roche is a global biopharmaceutical company that has used one-to-one marketing by developing targeted campaigns to reach patients with specific genetic profiles, using personalized messaging and digital tools to deliver relevant information and resources to those patients.

# Cloud Computing

## What is Cloud Computing?

Cloud computing is a technology where the software is not installed on individual desktops. Instead, it is hosted on a web-based server, offering on-demand software availability.

## Cloud Computing in Pharma

Early adopters of cloud computing among pharmaceutical companies, such as Pfizer, Johnson & Johnson, and Eli Lilly, used Amazon Web Services (AWS) and Amazon EC2 (Elastic Computer Cloud) in many areas like R&D, process proteomics, bioinformatics, statistics, and adaptive trial design to improve performance and reduce costs. While Pharma has been relatively slow in adopting cloud technology due to some security concerns initially, the subsequent improvisation in security features and the COVID-19 pandemic are driving the companies to embrace the cloud like never before.

According to Healthcare Information and Management Society analytics survey report, the adoption rate of cloud computing in the pharma industry has increased significantly, with over eighty percent of companies leveraging cloud services. Lower infrastructure costs, efficient management of workloads, and improved business agility are some of the main reasons for this increasing cloud adoption.

Cloud is more than just a storage solution. Leading biopharmaceutical companies increasingly recognize that the cloud can help them shrink innovation cycles, standardize processes across global operations, scale up rapidly, and enable analytics, among other benefits.

Over eighty percent of pharmaceutical companies say they apply cloud technology in many areas across the value chain, such as research and early development, market access, and commercial and medical applications.

Furthermore, we can see the transformative nature of cloud technology from the fact that Moderna delivered its first clinical batch of its COVID-19 vaccine candidate to the US National Institutes of Health for Phase I trial only 42 days after the initial sequencing of the virus using cloud technology. So how did Moderna do it? According to Stéphane Bancel, Moderna CEO, *you don't have to reinvent anything. You just fly*!

### What Pharma Leaders Say About Cloud

Andrea Del Miglio, Tobias Hlavka, Jeffrey Lewis, Manuel Möller, and Pablo Prieto-Munoz highlight the biggest advantage cloud technology can offer to pharmaceutical companies in their article, *The Case for Cloud in Life Sciences*, published in McKinsey.com on October 28, 2021. Here's what they wrote: *Cloud technology can help life sciences companies deliver brand-new services in less than six months, increase release frequency from quarterly to weekly, and slash deployment-lead times from days to hours*.

In addition, they quote what some of the big pharma leaders said about what Cloud technology has done and could do to the pharma industry. Consider these, for example:

- Jim Swanson, Chief Information Officer at Johnson & Johnson: *Ubiquitous data, together with elastic, scalable cloud and edge computing, is part of what's enabling us to innovate and scale in such unprecedented ways*.
- Christopher Weber, CEO of Takeda**:** *My vision is to deliver transformative therapies and better experiences to patients, physicians, and payers faster than previously possible*.

### How Cloud Computing Supports Pharma Operations

Cloud computing helps businesses save on software and hardware costs essential for basic marketing infrastructure costs. Further, it helps pharmaceutical companies meet HIPAA (Health Insurance Portability and Accountability ACT) compliance requirements. Mike Cox explains how cloud computing can support your marketing operations in Pharma in an article titled 8 *Ways Cloud Computing*

*Supports Pharma Marketing Operations* published in P360: Powered Possibilities blog on October 2, 2019. Here is the gist:

Pharma marketers must have direct access to their marketing and audience data anytime, anywhere to maintain a strong marketing strategy and support each campaign from the ground up. Cloud computing makes this possible in eight ways:

1. **Access to Multi-Device Support.** Pharmaceutical industry generates petabytes of data every second via multiple data streams, like R&D and CRM systems. It's a big challenge to leverage this humongous data regarding where to store it and how to access it in time. Cloud computing allows marketers to store important data in a global network that you can access anywhere and on any device. Thus, it allows pharma marketers to function more efficiently where they are instead of having to be in the office to access specific data. Furthermore, it allows for the simple exchange of information and helps promote clinical trial recruitment. It also helps patients access their medical records and to contact their physicians.
2. **Direct Access to Huge Patient Data Instantly**. Cloud technology helps pharma marketers easily access and share relevant files from anywhere, improving collaboration and communication between teams and clients. Also, with patient data pouring in from everywhere, including social media, apps, fitness wearables, and the internet, cloud computing ensures pharma marketers access that data anytime from anywhere. According to a patient survey, more than 85 percent of patients said they were confident in their ability to take responsibility for accessing online resources to help them do so.
3. **Maximize Cost Efficiency**. Contrary to the popular notion, implementing cloud computing into pharma marketing operations is much more affordable than traditional methods. This is because you don't have to spend much money and time on infrastructure, software, and hardware. The cloud is already out there, ready and waiting for your data. All you need to do to get started is to pay the initial sign-up fee for the subscription services of your choice. In addition, cloud

computing helps you reduce your total operational costs and improve process efficiencies. It also creates an open communication channel to facilitate faster decision-making and more streamlined patient engagement and market access processes.

4. **Increased Security**. In the past, security was a problem with cloud technology and was a primary reason pharmaceutical companies would stay far from it. However, recent technological advancements in encrypting data added multiple layers of security in cloud computing and improved it significantly. As a result, data such as contacts, sales tracking, patient information, content, and strategies are heavily protected by the cloud, ensuring that your marketing operations remain safe and secure.
5. **Increased Mobility**. We are living in a mobile world, marketing more so. Further, smartphones are used for much more than just making phone calls or texting. Cloud technology ensures that your marketing operations keep up with the continuously changing and heavily competitive pharma market by helping your team access information wherever they are and whenever they need it.
6. **Improved Analytics**. Advanced cloud analytics can help track and better guide patients through their journey and help marketers understand what matters most to their audience. For example, analytics through the cloud can track where the visitors have clicked, what they have done during their visits, and how their experience has been so that you can serve specific content relevant to those patients to guide them further in their journey.
7. **Improved Communication**. Patients, HCPs, hospitals, and third-party companies need to know that what they buy will solve their problems before spending any money. Through the cloud, the BPaaS (Business Process as a Service) customer service system ensures communication is continuously flowing, addressing their needs. BPaaS system can be a major game-changer for your pharma marketing operations as it collects

important information related to patient data, flags common problems, categorizes client service histories, and quickly shares, filters, and pushes solutions across various channels.

8. **Optimizes Digital Media Spend**. The cloud-based BPaaS system can assist the marketing department in managing its marketing strategy on social media to make better decisions for spending effectiveness and ROI based on campaigning on certain platforms by finding timely answers to questions like:

- How much do you spend on social media posts?
- Where do you focus these posts geographically?
- How do you know how much to divide your posts among all the platforms?

Finally, cloud computing technology can help pharma marketers have more time for other tasks that need their attention, see actual results from their work, and feel more confident going forward with streamlined marketing operations.

CASE

20

# Dr. Reddy's Laboratories Embraces Cloud Technology!

Dr. Reddy's, a leading international pharmaceutical company from India, moved its Active Pharmaceutical Ingredients (API) business to the cloud to improve visibility into its deals pipeline, track its end-to-end sales cycle, and better analyze its sales funnel.

As Rajdeep Ghosh, Head of the Digital Centre of Excellence at Dr. Reddy's Labs, says, *Today's businesses realize the power of digital technologies and platforms. So instead of reinventing and building from scratch, you can leverage Cloud services and create immense value for your business. And to migrate to cloud technology, you don't have to spend millions of dollars. With open source technologies and the Cloud Platform & Services, we don't have to build everything from scratch. Today's technology landscape allows you to drive a frugal innovation approach to digital transformation. Cloud Providers offer Today a basket of ready-to-use services you can conceptualize, build, and integrate to release any solution faster.*

Further, Rajdeep Ghosh added that the key here is to look at Cloud as PaaS (Platform As A Service) and extract value, not just as another alternative to on-premise infrastructure. Using Cloud services, Dr. Reddy's have reduced the overall development timelines by 30 percent. Dr. Reddy's have been leveraging Cloud Services such as RDS (Remote Desktop Services), Redis Cache (In-memory data store based on Redis Software), S-3 Bucket (Simple Storage

Service available from Amazon Web Services), API Gateway (API management tool that sits between a client and a collection of backend services), Cloud Front (Globally distributed content delivery network offered by Amazon Web Services), Sage Maker (Cloud Machine Learning Platform), SNS (Simple Notification Service), Lambda (Serverless, event-driven computer service that lets you run code for virtually any type of application from over 200 Amazon Web Services and SaaS applications, and only pay for what you use), Cloud Watch (Monitoring tool of Amazon Web Services and the applications you run in real-time), and Pub/Sub (Publish/Subscribe messaging is an asynchronous service-to-service communication method used in serverless and micro-services architectures).

Bodhtree, an IT consulting firm, helped Dr. Reddy's implement Salesforce CRM, using its MIDAS data integration engine to integrate with existing Mail, Calendar, and SAP systems. As a result, Dr. Reddy's plans to increase its revenue by streamlining its sales business on Salesforce by more than 30 percent.

Speaking about implementing the Salesforce CRM, B.V.Srinivas, Deputy CIO (Chief Information Officer) at Dr. Reddy's Laboratories, said: *We eliminated bottlenecks within the sales cycle almost immediately because of the greater visibility we gained into sales performance. In addition, Salesforce's ability to seamlessly integrate data captured offline with online systems is of immense help to our field sales team, who need to track and update customer data regularly but do not have internet connectivity at all times. Salesforce delivers all of the benefits of a world-class CRM system within a cost-effective 'pay as you go' model, which perfectly fits Dr. Reddy's 'quality without the cost' ethos.*

Lee Thomson, Senior Vice President of Asia Pacific at Salesforce, commented on Dr. Reddy's embracing of Salesforce's Cloud-based CRM. He said: *Dr. Reddy's is a prime example of cloud integration adding value to legacy systems. With cloud computing, there are no hidden costs in terms of hardware, software, or 22 percent annual maintenance fees. You know exactly what you're going to pay for, and people love using Salesforce CRM.*

(Source: Adapted from the articles—1. Vishal Krishna, Why Dr. Reddy's is Thinking Like A Tech Company As It Focuses on Digital Transformation? Published in YourStory blog on December 4, 2019, and 2. Bodhtree Press Release, Dr. Reddy's Laboratories Reengineers Sales Process in the Cloud with Salesforce, PR.com, February 10, 2011)

## AR (Augmented Reality) and VR (Virtual Reality) in Pharma Marketing

What do Pokémon Go and Excedrin have in common? Both are using Augmented Reality (AR) and Virtual Reality (VR) to persuade consumers to be more active (Pokémon Go) or to be more compassionate (Excedrin). However, Pokémon Go is the revolutionary Augmented Reality-based mobile game that made the underlying technology extremely popular and accessible. In contrast, GlaxoSmithKline's branded Excedrin Migraine Experience is the world's first Virtual Reality (VR)-based migraine simulation that provides a new understanding of migraine. The migraine experience is designed to allow non-sufferers to safely experience symptoms, including disorientation, aura, sensitivity to light, and blurred vision.

### What is Augmented Reality (AR)?

Thomas Caudell, a researcher at Boeing, coined 'Augmented Reality' (AR) in 1990 to describe how the head-mounted displays that electricians used when assembling complicated wiring harnesses worked. However, Augmented Reality became the news when Google Glass was tested in 2013.

Techopedia, the well-known IT education website, describes Augmented Reality (AR) as *an interactive experience of a real-world environment where objects in the real world are enhanced by computer-generated perceptual information, sometimes across multiple sensory modalities, including visual and auditory, haptic, somatosensory, and olfactory*. Augmented Reality thus combines real and computer-based scenes and images to deliver a unified but enhanced view of the world.

Augmented Reality employs computerized simulation and techniques such as image and speech recognition, animation, head-mounted, hand-held devices, and powered display environments to add a virtual display on top of real images and surroundings.

### What is Virtual Reality (VR)?

Virtual Reality (VR) is a three-dimensional, computer-generated environment where you can explore and interact. It completely replaces the user's world environment with a simulated virtual Reality. The person experiencing virtual Reality becomes a part of this virtual world or immersed within this environment and can manipulate or perform a series of actions while he is there.

Virtual Reality (VR) uses computer technology to create a simulated environment. Unlike traditional user interfaces, VR places the user inside an experience. Instead of viewing a screen in front of them, users are immersed and interact with three-dimensional worlds. By simulating as many senses as possible, such as vision, hearing, touch, and even smell, the computer is transformed into a gatekeeper to this artificial (virtual) world.

Virtual Reality creates this immersive experience with the help of devices such as HTC Vive or Oculus Rift. These devices help transport the users into several real-world and imagined environments, such as the middle of a squawking Penguin colony or the back of a dragon, or even into a bloodstream, respiratory tract, or other physiological processes like medical and pharmaceutical environments.

### Difference Between AR and VR

Augmented Reality (AR) enhances the real-life environment with some computer-generated elements. So the virtual scenes are mixed with the real ones, overlaying the digital details in the real world. In addition, Augmented Reality often adds digital elements to a live view by using the camera on a smartphone. Snapchat lenses and the Pokémon Go game are some notable examples of Augmented Reality.

Virtual Reality (VR), on the other hand, completely replaces the real world with a virtual environment. That means everything you see looks quite life-like, though, in fact, it is just a simulation.

You must use a head-mounted controller or glasses to dive into the Virtual Reality world. At the same time, Augmented Reality mostly uses commonplace devices like tablets, smartphones, or laptops. Of course, Augmented Reality can also be applied through Microsoft HoloLens, which is currently very expensive to be widespread.

## Why AR or VR?

Why do marketers, in general, and pharma marketers, in particular, need AR or VR? Are they really necessary? Here are five compelling reasons for making AR and VR an integral part of your marketing strategy. VR sessions can increase the quality and attractiveness of visits by medical representatives.

1. Modern VR builds upon mechanisms of action (MOA) animations, which have become a mainstay in pharmaceutical marketing. For example, today, no blockbuster drug gets released without a 3-D animation. While MOA provides clear and concise information about a drug, they pale compared to the experience of being in the bloodstream or penetrating a cell wall to deliver a drug, which only Virtual Reality can create.
2. Both AR and VR significantly improve patient-HCP communications, which results in better outcomes. As Drew Griffin, senior technical architect at Razorfish Health, said: *Given that patient-HCP communication can be qualitatively dense, any technology such as AR/VR can provide just-in-time, easy-to-digest, context-sensitive data will be an unprecedented and invaluable tool to educate patients during diagnosis and treatment of their conditions. This allows pharmaceutical companies to reach healthcare professionals and patients with their creative communication.*
3. With most healthcare professionals (HCPs) and patients using the web to access health, medical, or prescription drug information, we are in the middle of an ocean full of digital health information. How can you stand out in such a scenario? Creating visual and educational content without being ad-driven can be highly interesting and involving. Augmented Reality and Virtual Reality can help us achieve these communication objectives as they can create immersive experiences.
4. In the US, less than 50 percent of the pharmaceutical sales force has access to physicians. Still, US drug companies spend up to $5 billion each year sending their representatives to meet doctors. In 2015 alone, it was reported that pharmaceutical companies in the US wasted 1 to 1.5 billion dollars on marketing

by making infeasible calls to physicians. The Pharma industry needs an effective communication tool to extend its reach to reluctant physicians. AR and VR are probably excellent means to achieve this.

## Augmented or Virtual Reality?

Several marketing situations are ideally suited for implementing Augmented Reality or Virtual Reality. As these technologies become more affordable, knowing when and how to deploy them most effectively is crucial. The core philosophy for AR and VR revolves around empathy. Both technologies are experiential. However, when you have decided to use these technologies, the first critical decision is: whether to use AR or VR? Besides their technical differences, these approaches have different applications in marketing. Here are two broad distinctions between AR and VR:

A. Virtual Reality (VR) is typically an enclosed environment enabling an immersive experience. Therefore, one can use Virtual Reality to let healthcare providers and caregivers experience another person's symptoms, providing a more definite sense of how a disease affects their lives. On the other hand, Augmented Reality (AR) often involves projecting onto an external surface and providing a shared experience. Augmented Reality is a good fit for educational campaigns. It can also demonstrate a novel therapeutic approach and a new drug's mode of action.

B. Like any other marketing technology, knowing when not to use Augmented Reality and Virtual Reality is essential. These new technologies' 3-D functionality and immersive nature should be a supplement, and they do not replace traditional approaches. When a print ad accomplishes the campaign's goal, using a VR or AR version does not make sense because they are more appealing and advanced. A good rule of thumb is that VR and AR are not the choices if there is no pressing need for empathy or hands-on education.

Augmented Reality (AR) and Virtual Reality (VR) are useful in pharmaceutical marketing in many ways. Consider these, for example:

- Augmented Reality (AR) can provide information about a drug or medical device visually and interactively. For example, an AR app could be used to show a 3D model of a drug molecule or to demonstrate how a medical device works.
- Another use of AR is to provide patient education. For example, an AR app could provide information about a disease or condition or help patients understand how to use medication properly.
- VR (Virtual Reality) can be used similarly, offering a more immersive experience. For example, a VR app could simulate a surgical procedure or provide a virtual tour of a pharmaceutical manufacturing facility.
- AR (Augmented Reality) and VR (Virtual Reality) can enhance the sales process. For example, an AR or VR app could be used to provide product demonstrations or simulate the use of a product.

Overall, AR and VR can be valuable tools for pharmaceutical companies to educate patients and healthcare providers about their products and to enhance the customer experience.

## Blockchain in Pharma Marketing

### What is Blockchain?

Blockchain is a list of records—an open and distributed ledger. As new records (blocks) are added to the list (chain), they're verified cryptographically across the network, maintaining the blockchain. As a result, blockchain is inherently secure and allows for transparent consensus without putting any central party in charge. Furthermore, blockchain technology can fix many of the vulnerabilities of the conventional supply chain in any industry. The advantage of a blockchain-based system is that competitors can collaborate on a shared platform without sharing sensitive information. Companies can safely work together in a shared, permanent ledger without giving up control or revealing their data. A single company does not own the ledger in the blockchain system. Instead, it is governed by all members of a network. That is how the blockchain delegates the work of checks and balances to cryptography and code and helps reduce friction, expose fraud, and assure product authenticity with great speed.

### The Pandemic Exposes Supply Chain Inefficiencies

The recent global pandemic, COVID-19, exposed the inefficiencies of the pharmaceutical and medical supply chains worldwide. For example, leading hospitals in almost all countries struggled with shortages of basic safety equipment such as masks, gowns, face shields, and sanitizers during the pandemic. The hospital authorities and government officials were not ready to meet the swift and sudden escalation in demand, bringing to light the long-festering problems in the medical supply chain. In the US, CDC and FDA officials were concerned that fraudulent respirators were circulating and fraudulent test kits for COVID-19 were being sold online. As a result, the Federal government in the US has placed more than $110 million in N95 mask orders at high prices with unproven mask vendors.

Blockchain has been a disruptive technology for a decade, and many industries have embraced it. However, when the COVID-19 pandemic struck, the pharmaceutical industry realized that its global supply chain lacked connectivity and data exchange and that only blockchain could help resolve it.

When China's manufacturing system almost reached a grinding halt, global pharmaceutical supply chain cracks started to appear. That's when the healthcare industry realized the urgency to improve the global supply chains worldwide that started crumbling when borders worldwide were closed, followed by the reduced workforce, lockdowns, and an insatiable demand growing for goods and services internationally.

### How Blockchain Can Help Pharma

Arun Ghosh, a KPMG analyst, says that the primary use for blockchain in the pharmaceutical industry is to serve as a ledger of truth for sharing complex information with regulators, pharmacy benefit managers, contract manufacturers, physicians, patients, academic researchers, and R&D collaborators, and others. Furthermore, blockchain can improve nearly every component of the pharmaceutical value chain, from increasing supply chain transparency to enhancing data collection in clinical trials and optimizing marketing and advertising expenditure.

Blockchain is crucial for pharma as it provides traceability throughout the product lifecycle by relating data, propagating it, and distributing it securely throughout the organization. In addition, the pharma industry requires transparency in its clinical trials, supply chain control as it scales up, product validation, and quality assurance in the approval process and commercialization. But, perhaps, the most important advantage that blockchain technology can offer the pharmaceutical industry is to help solve the long-festering problem of counterfeit drugs.

Blockchain technology has the potential to help with many aspects of pharmaceutical marketing. Consider these, for example:

1. *Supply Chain Traceability*: Blockchain can create a secure and transparent record of a product's journey from the manufacturer

to the end consumer, which can help prevent counterfeit drugs from entering the supply chain.

2. *Data Integrity*: Blockchain can be used to ensure that data about a product, such as clinical trial results, is accurate and cannot be tampered with. This can help to build trust in the product and increase patient confidence.
3. *Compliance*: Blockchain can automate compliance with regulations, such as the FDA's Drug Supply Chain Security Act (DSCA), by providing a tamper-proof record of a product's journey through the supply chain.
4. *Direct-to-Consumer Marketing*: Blockchain can be used to create a secure and transparent way for patients to access information about products and manage their health data.
5. *Patient Engagement*: Blockchain can be used to create a secure, decentralized platform for patients to share their health data with healthcare providers, researchers, and pharmaceutical companies. This can help to improve patient engagement and lead to more personalized treatment options.

Thus, blockchain technology can help improve the transparency, security, and efficiency of the pharmaceutical supply chain and marketing process, ultimately benefiting patients and companies.

## Cybersecurity in Pharma Marketing

Cybersecurity is crucial in the pharmaceutical industry due to the sensitive nature of the information being handled, such as personal health information, financial data, and intellectual property. Therefore, a cyber-attack on a pharmaceutical company can have severe consequences, including the loss of confidential information, the interruption of production and supply chain, and damage to the company's reputation.

Here are a few reasons why cybersecurity is important in pharmaceutical marketing:

1. *Data Protection*: Pharmaceutical companies handle sensitive information, such as patient data, clinical trial results, and confidential business information. Cybersecurity measures are necessary to protect this data from unauthorized access, theft, or alteration.
2. *Brand Protection*: A cyber-attack on a pharmaceutical company can damage the company's reputation and lead to a loss of trust from customers, healthcare professionals, and regulatory bodies.
3. *Compliance*: Many countries have regulations to protect sensitive information in the healthcare industry. Pharmaceutical companies must comply with these regulations, and cybersecurity measures are necessary to meet these requirements.
4. *Supply Chain Security*: Cybersecurity is important in the pharmaceutical supply chain to prevent counterfeit drugs from entering the market and to ensure the integrity of the supply chain.

Therefore, cybersecurity is essential to protect the sensitive information of pharmaceutical companies, maintain brand trust, comply with regulations and ensure the integrity of the supply chain to ensure the safety and well-being of the patients.

# Patient Engagement

## Patient Centricity

Patient centricity is a philosophy and approach that places the needs and preferences of patients at the center of the healthcare system. It is a holistic approach that considers the physical, emotional, social, and economic aspects of a patient's well-being.

A patient-centric approach involves healthcare providers, pharmaceutical companies, and other stakeholders working together to understand and address the unique needs of individual patients. This includes involving patients in their healthcare decision-making process, providing education and support, and offering personalized treatment options. Thus, the primary goal of patient-centricity is to improve the patient experience, increase patient satisfaction, and ultimately improve health outcomes.

Involving patients in developing treatment plans, providing them with the education and tools they need to make informed decisions about their health, and ensuring that their care is tailored to their needs are essential to achieving patient centricity.

Patient-centricity in the pharmaceutical industry means that drug development and marketing efforts are focused on addressing the specific needs of patients and that the industry works to make drugs more affordable and accessible to patients.

## What Pharma Should Do to Become Patient-Centric

Firstly, pharmaceutical companies must realize that becoming a patient-centric organization is an ongoing process that requires commitment and willingness to adapt to the needs and preferences of patients and to strive to improve the health outcomes of patients continuously. In addition, pharma companies can take the following steps to become patient-centric.

1. Involve patients in the drug development process so you can consider their needs and preferences and design new treatments accordingly.

2. Enable patients by educating and supporting them with information about products, side effects, their health, treatment options, and potential outcomes.
3. Offer personalized treatment options considering a patient's unique needs, preferences, and medical history.
4. Ensure patients access the necessary treatments by making drugs more affordable and accessible.
5. Engage with patients through various channels, such as social media, patient advocacy groups, and other forums, to understand their needs, pain points, and preferences and respond to their questions and concerns.
6. Respect patient privacy and security by adhering to strict data protection regulations and guidelines, and be transparent about how their data is collected and used.
7. Collaborate with other stakeholders, such as healthcare providers, payers, and patient advocacy groups, to improve the patient experience and health outcomes.
8. Continuously measure and improve the patient experience through surveys, feedback, and other methods to identify areas for improvement and make necessary changes.

**Patient Engagement**

Pharma companies can become patient-centric only when they engage with patients effectively throughout their journey. There is more to patient engagement than providing information and education about products, services, health conditions, treatment options, and potential health outcomes. It can take many forms. For example, pharma companies can:

1. Involve patients in the drug development process early to design treatments that address patients' unmet needs and concerns.
2. Provide patient education and support so they understand their health condition better, and how to manage it.
3. Empower patients to manage their health through self-management programs and tools.

4. Encourage patient communication. Patients should be encouraged to communicate with their HCPs, ask questions and provide feedback.
5. Engage patients digitally by providing digital tools to help them engage with their HCPs.
6. Partner with patient advocacy groups. For example, a pharmaceutical company can partner with a patient advocacy group to co-develop a patient education program involving patients in creating the content.
7. Conduct patient research, involving them as partners and not just as subjects. This means involving them in the design, implementation, and dissemination of research findings.
8. Create patient-centered care by treating them with respect and empathy, considering their unique needs and preferences in care delivery.
9. Actively listen to patients and consider their feedback.

### How Pharma is Engaging Patients

Many pharmaceutical companies have implemented patient engagement strategies. Other companies can emulate some of their strategies. Consider the following, for example:

1. **Novo Nordisk:** Novo Nordisk, a global pharma major specializing in diabetes care, has implemented a patient engagement strategy that includes providing patient education and support through various channels, such as in-person events, digital resources, and online communities. They also involve patients in developing their products and services through patient advisory boards and engagement groups.
2. **Pfizer:** A global pharmaceutical leader, Pfizer's patient engagement strategy includes providing patient education and support, involving patients in research studies, and utilizing digital platforms to engage with patients. In addition, Pfizer has a patient engagement program, *Pfizer Patient Assistance*, which provides financial assistance to eligible patients who need help paying for their Pfizer prescriptions to improve access.

3. **Sanofi:** Another global pharmaceutical major, Sanofi, has implemented a patient engagement program that includes the development of new treatments, providing education and support, and utilizing digital platforms to connect with patients and engage them. They also have a patient engagement program, *Sanofi Patient Connection*, which provides financial assistance to eligible patients who need help paying for their Sanofi prescriptions to improve access.
4. **Janssen:** Janssen, a pharmaceutical division of Johnson & Johnson, has implemented a patient engagement strategy that involves patients in developing new treatments, providing patient education and support, and utilizing digital platforms to engage with patients. Their patient engagement program, *Janssen CarePath*, provides support and resources to help eligible patients access Janssen's prescribed treatments and therapies.
5. **AstraZeneca**: AstraZeneca's, a global biopharmaceutical company, patient engagement strategy includes involving patients in developing new treatments, providing patient education and support programs, and utilizing digital platforms to engage with patients. AstraZeneca's Patient Assistance Program provides financial assistance to patients who need help paying for their AstraZeneca prescriptions.
6. **Merck:** Merck, a global pharma major from the US, has a patient-centric approach that includes developing patient assistance programs, providing financial assistance to improve access to their medications, and partnering with patient advocacy and advisory groups. They also conduct patient-focused research to understand the patient experience better and improve their products and services.
7. **GlaxoSmithKline:** GlaxoSmithKline's patient-centric approach includes patient engagement programs covering a range of activities, including developing patient assistance programs and providing educational resources and financial assistance for eligible patients to improve access to their medicines. The company also partners with patient advocacy groups and

engages patient and caregiver groups in clinical trial design to gain feedback on new drugs.

8. **UCB:** UCB is a global pharmaceutical company focusing on developing and commercializing treatments for diseases of the nervous system and immunology. The company has implemented several patient engagement strategies to understand better and meet patients' needs. For example, UCB involves patients in developing new treatments by gathering input from patient advocacy groups and conducting patient research studies. Like leading pharma companies, UCB provides patient education support and engages them on digital channels. Furthermore, UCB is committed to ensuring that its medicines are accessible to the patients who need them, offering a range of support programs and initiatives to help ensure access, affordability, and responsible pricing. Also, UCB continuously measures and improves the patient experience through surveys, feedback, and other methods to identify areas for improvement and make necessary changes.
9. **Cipla**: Cipla, one of the leading international specialty pharmaceutical companies, has a patient engagement strategy that includes patient education, patient support, and engagement through various digital channels and mobile apps. Cipla's mission is to make medicines affordable to as many people as possible with patient assistance programs and responsible pricing. Also, Cipla engages patients in their research and development activities.
10. **Sun Pharma**: Sun Pharma, the fourth largest international generic company, has implemented patient engagement strategies, such as patient education and support programs to improve access and affordability, collaboration with patient organizations, and digitally engaging with patients. In addition, Sun Pharma engages patients in their research and development activities by including them in the design and execution of clinical trials and gathering patient input on drug development.

The following case shows how a patient engagement strategy by Cipla helped asthma patients in India improve compliance to inhalation therapy.

CASE

21

# Cipla's Patient Engagement Strategy!

Cipla, an international specialty company from India, launched Asthalin Dry Powder Inhaler (Salbutamol) in 1988 and appointed paramedics outside the chest specialist's chamber to educate patients on how to use the inhaler. Whenever the chest specialist used to prescribe Asthalin DPI (Dry Powder Inhaler), Cipla's paramedics' team members used to educate the patient about how to use the DPI and why it is important to use it as per the doctor's advice to get relief from their breathing trouble which is caused by asthma.This revolutionized asthma treatment at that time. Earlier, only oral solids, syrups, injectables, and antibiotics were used to treat asthma, but since 1988, slowly, over time, patients have started using inhalation therapy.

Even though MDI (Metered Dose Inhalation) is proven superior and scientifically more suitable to DPI (Dry Powder Inhalation), DPI became more popular than MDI in India due to Cipla's patient education and engagement strategy. Thus, Cipla's patient engagement strategy helped doctors to treat their patients better and improved patient compliance.

Patient engagement is crucial because it helps to improve the patient experience and health outcomes, increase patient satisfaction, and helps to build trust between patients and healthcare providers, including pharmaceutical companies.

## Phygital: The Future of Engagement

### What is Phygital?

Chris Weil, Chairman-CEO at Momentum Worldwide, an international practice in consulting, strategy, analytics, and experience design, first coined the word — Phygital in 2007 to describe the inseparable connection between the physical and digital worlds. Phygital is derived from the contraction of two words (Physical and Digital) into a single word (Phygital). Phygital is using technology to bridge the digital and physical worlds to provide a unique interactive user experience.

The physical and digital boundaries have been blurred for some time, creating a seamless omnichannel experience in the retail, transportation, hospitality, and entertainment industries. Successful companies today do not follow a physical or digital strategy but a united phygital strategy that unites both sides to create an immersive digital experience for the customer.

The phygital trend accelerated during the COVID-19 pandemic, as life moved online, and brands, companies, and work itself had to pivot to digital channels quickly.

Blake Morgan, a globally recognized keynote speaker and the author of the bestselling book, The Customer of the Future, wrote in an article, *The Rise of the Phygital Experience*, published on forbes.com on May 16, 2022, that *Marketers must make every interaction of their brands with customers count and unite what customers want most in a digital channel and what they want most in a physical experience to create a united phygital experience*.

How do you create a strong phygital experience? Like all aspects of customer experience, creating a great customer experience starts by understanding your customers, what matters to them most, and how they like to acquire and interact with that. A quality phygital experience understands customers' needs and provides excellent digital tools and in-person experiences to move between channels seamlessly.

## Pharma and Physical Marketing

Traditionally, pharma marketing relied heavily on physical marketing and on — in-person channels — mainly sales reps, physical meetings, and congresses — pushing messages about their products and services. However, the pharma marketing landscape has been changing over the past few years, and the pandemic has increased the speed of change, causing many challenges. Here are the main reasons and challenges.

- Digital natives among HCPs (Healthcare Professionals) are increasing, with the tipping point (51 percent) being reached in 2014. Digital native physicians are estimated to be about 75 percent of total physicians by 2030. These physicians are digitally savvy, prefer digital channels, and want to receive relevant communication conveniently on the channels they want.
- Pharma marketers in the off-patent space find it difficult to differentiate their products and continue to stick to sales pitches and push messages. As a result, initiatives beyond the pill are missing.
- Pharma companies cannot gain adequate attention from doctors as they are unwilling to spend more than a few minutes with sales representatives as they are not meeting their information requirements.
- Unable to provide compelling reasons for physicians to prescribe their products, many pharma companies in the branded-generics markets are resorting to the soft option of transactional marketing. However, with some rules and guidelines in place, it is also becoming difficult to provide upfront benefits to HCPs on a *quid pro quo* basis. Instead, it forces the industry to move to a relationship-based approach to engage physicians with persuasive, evidence-based communication.
- Even offline, physical marketing events such as seminars and conferences through key opinion leaders are becoming difficult for pharma companies to engage physicians.

- Empowered patients are demanding seamless customer experiences that they are used to from other sectors of healthcare too.

## Digital is the Way Forward

A steady decline in the quantity (reducing access to physicians) and quality of engagement and an increasing adoption rate of physicians to digital channels have been pushing pharmaceutical companies to use digital channels for improving physician engagement.

When the digital revolution is changing doctors' behavior in seeking information, can pharma continue with its communication strategies and not go forward with digital ways?

Furthermore, a Docplexus study on understanding what information HCPs want from pharma and how they want it revealed that:

- 79 percent of physicians are unable to get full information during visits from medical representatives
- 75 percent of physicians go online for more information
- The preferred schedule of physicians for online communication is once a week between 1 to 4 PM or 7 and 10 PM.
- As regards the types of information, 42 percent of HCPs want medical types, with 35 percent looking for new treatment regulations, and 23 percent wanting to know the services by pharma.

## Common Multichannel Pitfalls

There are some common misconceptions about digital, and marketers will do better by avoiding these:

- Most people believe that multichannel is digital only. It is not.
- Some marketers think and use digital only as a playground and use it for some pilots and small projects. However, Digital is not a playground but a mainstream.
- Do not equate digital with innovation.
- Do not place digital in isolation. Make it an integral part of your strategy.

- Do not see digital only as a way to enhance sales rep engagement with the target group. Instead, think of a much larger role for Digital in your total strategy.
- Remember, Digital is not a substitute for face-to-face communication. Use it right, and it can augment your overall strategy significantly.

## Is Phygital the New Omnichannel?

Phygital blurs the line between physical and digital channels to create a consistent omnichannel experience. There is no longer a separate physical and digital strategy but an integrated phygital strategy that unites physical and digital sides to create an immersive customer experience no matter which channel they access for information—product, or service.

Every multichannel or omnichannel strategy should contain a balanced mix of the three media types—Owned, Paid, and Earned. Here is a brief account of each media type, what it is, its role, and its benefits:

- **Owned Media** comprises those channels that the company fully controls. For example, a website, mobile site, blog, newsletter, and press release form part of the owned media. Owned Media helps build longer-term relationships and establish a continuous presence. The advantages of owned media are— full control, cost efficiency, longevity, versatility, and addressing different target audiences. The downside of owned Media is that it often lacks trust among HCPs, the source being the company or the brand, and their perceived vested interests.
- **Paid Media** is the presence of your company acquired through buying space on these channels. Display advertising, SEA (Search Engine Advertising), and sponsorships are examples of paid media. Paid Media is used essentially to drive traffic to your owned media. The advantages of Paid Media are that it gives you control over costs and immediate impact and is easily scalable. However, the response rates with Paid Media are declining, and banner blindness is increasing.

- **Earned Media** is when your customers become the channel. Notable examples of Earned Media are Word-of-Mouth (WoM), Buzz and Viral marketing, and Retweets and Likes on social media platforms. Earned Media is very important in establishing stronger credibility and improving engagement. However, the company does not have any control over Earned Media. Also, it is hard to measure and scale up.

Here is a table giving you an illustrative list of the various channels available to engage your customers.

**Table 6.12 The Phygital Channel Mix (Offline and OnLine)**

| Offline or Physical Channels | Online or Digital Channels |
|---|---|
| • Promotional call<br>• Direct Mail<br>• Newsletter<br>• Sales rep visit<br>• Sales rep tablet detailing<br>• Samples<br>• Promotional SMS<br>• Local scientific meeting<br>• International scientific meeting<br>• Medical Journal<br>• CME<br>• Congress Booth for displaying company's products<br>• MSL (Medical Science Liaison)<br>• KOL meeting<br>• Medical Education<br>• Phase IV Clinical Trail | • eDetailing<br>• eNewsletter<br>• Remote Engagement<br>• Website<br>• eMSL<br>• Webcast<br>• eCME<br>• Smartphone App<br>• KOL webinar |

In sum, Owned Media enables you to engage with customers, Paid Media with strangers, and Earned Media with advocates. Therefore, you must use a balanced mix to effectively reach and engage your target audience.

### Implementing Phygital

Now that we know the importance of phygital marketing, how do we implement or practice it? Firstly develop a phygital strategy by understanding the target audience thoroughly regarding their needs, wants, and channel preferences and a convenient time for interacting. Consider the following steps:

1. Develop a brand strategy covering all key aspects, such as brand objectives, positioning, business opportunities, and key stakeholders.
2. Translate the strategy into something actionable by identifying leverage points, behavioral objectives, and differentiated messages.
3. Decide on the conversion points— what content or services will drive the action and change we want?
4. What are the ideal channels to achieve this—physical and digital?
5. Prioritize your action plan by deciding on what is feasible. Also, study the impact of these tactics.
6. Measure what matters. How can we measure the interactions and engagement?
7. Invest adequately in creating the necessary digital and physical infrastructure.
8. You must blend insights from offline and online behaviors into physician perception about the therapy area, your molecule, and your competition. It would help you develop compelling communication strategies for your brand.
9. Develop and create relevant, modular content that is channel-specific yet integrated to ensure uniformity and consistency of messaging across the channels to create a seamless experience.
10. Train your sales representatives thoroughly on scientific and technical aspects to build their capabilities for orchestrating multiple channels— physical and digital in effective physician engagement.
11. Implement closed-loop marketing to provide your sales representatives with the relevant next-best action tactics based on continuous measurement and monitoring of the impact of their interactions with HCPs.

12. Finally, creating holistic engagement with your audience through various channels is crucial to succeeding with phygital. An integrated approach will help you seamlessly improve the experience of your customers (HCPs) and consumers (patients). Studies show that good omnichannel engagement results in up to 90 percent retention of customers.

CHAPTER 7

# Transformational Marketing in Pharma: Two Case Studies

Transformation is not about improving. It's about rethinking.

— *Malcolm Gladwell*

# Transformational Marketing in Pharma : Two Case Studies

Transformational marketing occurs when you take monetary transactions and other unethical practices out of the relationship equation. While most pharmaceutical companies hesitate to even conceive that idea, GlaxoSmithKline implemented a revolutionary marketing strategy in the pharmaceutical industry did just that! GSK showed the way to the industry that it is possible to move from the current predominantly transactional mode of marketing that is rife with unethical practices to a transformational marketing model based on the solid foundation of scientific marketing, customer and patient-centric focus, improving loyalty, and restoring Pharma's declining reputation to its earlier levels of respect and glory!

EyeForPharma published a highly insightful white paper—*GSK's new ethical approach: Is it delivering?* It is a review, analysis, and key lessons— with contributions from Stewart Adkins, Marc Bacon, Christopher Bowe, John LaMattina, George Katzourakis, Matthew Smith, Murray Stewart, Jack Whelan, and Debra Whitman. I have borrowed from this white paper in the following case to show and share how GSK tried to travel the road not less traveled but never traveled in a bold, unprecedented move, showing how the industry can transform from the current transactional mode to the transformational way of marketing.

Here is the case describing how GlaxoSmithKline transformed its marketing model.

CASE

22

# Transactional to Transformational Marketing in Pharma— GSK Shows the Way!

## GSK's New Ethical Approach

Over the past fifteen years, big pharma has been facing serious challenges, such as dwindling product pipelines, escalating costs of clinical trials, declining productivity, and profitability, flagging levels of engagement with customers, already lower levels of reputation dropping further, and increasing prices at a time when payers are demanding more value from their health budgets.

At the same time, continuing technology revolution and growing consumerism in healthcare have prompted a traditionally conservative industry to seek innovative, even radical, solutions. How is pharma responding? Pharma is embracing and exploring patient-centricity to gain a better understanding of the needs of the end-users of its medicines and customer experience to overcome the growing challenges.

Some major pharmaceutical companies have asked even more fundamental questions, such as:

- What is pharma's brand?
- Why has its reputation been so poor when it delivers such breakthrough, life-saving and life-enhancing medicines?

Pharmaceutical companies find it extremely difficult to communicate the good they contribute to society due to the high levels of distrust created by many scandals involving bribery and other unethical marketing practices, including payments to healthcare professionals for prescribing more of their products, and with each scandal—whether price-gouging, bribery or unethical sales practices—the big bad pharma image is reinforced in the minds of all— patients, customers, and payers.

Basically, the relationship between doctors and pharmaceutical companies is mutually reinforcing, which can lead to improved patient outcomes. Dr. Martin Makary, Professor of Surgery at Johns Hopkins University and Professor of Health Policy Management at the Johns Hopkins Bloomberg School of Public Health, echoes a similar sentiment when he says: *Doctors can be better doctors when they partner with the pharmaceutical company. Likewise, pharmaceutical companies can benefit from a relationship with doctors to understand the concerns of patients we see as patient advocates*. However, the current relationship between physicians and pharmaceutical companies is far from ideal. Dr. Makary adds: *The relationship between pharma and doctors is polluted with conflicts of interest and is not good for healthcare. It's bad for patients when sales reps from pharmaceutical companies are paid based on the number of drugs they sell. Sales targets on a monthly basis make sense in retail stores; they make sense at a car dealership but not in hospitals. It's a dangerous trend. Patients need to be at the center of the healthcare system*.

**What Thought Leaders Think**

Professor Sir Michael Rawlins, Chair of the UK Medicines and Healthcare Products Regulatory Agency and previously Chair of the UK National Institute for Health and Care Excellence (NICE), says that *Doctors should not be a part of the marketing arm of the pharmaceutical industry and they shouldn't be paid for doing marketing. However, it's not black and white; for example, if I do a clinical trial on a product made by a pharmaceutical company, I'd want to present that at scientific meetings. Is that wrong? Is that promotional? We need to tease out the levels of the problem and respond accordingly to ensure the broad trust of the professions and the public is maintained and sustained*.

On sales incentives, Professor Rawlins adds, *It is a perverse incentive to make sales targets coupled to remuneration; it is much wiser for the interaction between the representatives and HCPs to be based on factual information, honesty, and open dialogue with the material publicly available*.

Commenting on the changing pharmaceutical marketing environment, Dr. Virginia Acha, who spent more than a decade in academia before becoming Executive Director of Research, Medical, and Innovation Policy at the Association of the British Pharmaceutical Industry (ABPI), says: *We have the challenge of establishing transparency regarding how and why we interact*. This is a new environment where companies and doctors have to consider what patients will think about our relationship, and she raises some important questions regarding this:

- What will they know about it, and how will it be reported?
- In these new worlds of both engaging with people via social media and being reported on through social media, do we need to think a little bit more deeply about whether 'the codes we have in place are still the right ones?
- Do we need to do something different?

On July 2, 2012, the United States Justice Department announced that GlaxoSmithKline agreed to plead guilty and to pay $3 billion, which is the largest healthcare fraud settlement in US history to resolve its criminal and civil liability arising from the company's unlawful promotion of certain prescription drugs, its failure to report certain safety data, and its civil liability for alleged false price reporting practices.

All these unethical practices predated the tenure of Andrew Witty, the CEO of GlaxoSmithKline (2008-2017), who has been trying to improve the company's reputation. One of the initiatives he undertook to restore the company's reputation was pushing forward with efforts to develop medicines for developing nations, including a malaria vaccine that Glaxo is developing with the Bill & Melinda Gates Foundation.

Also, in an unprecedented move, Andrew Witty gave a full-throated apology in the press release disclosing the company's settlement with the Justice Department, saying. *While these originate in a different era for the company, they cannot and will not be ignored. On behalf of GSK, I want to express our regret and reiterate that we have learned from our mistakes*.

Andrew Witty did not stop with these. In addition, he has made the boldest move to transform the company's sales and marketing model with a transparent and ethical approach to engaging physicians and pharma's key stakeholders, which is unprecedented.

The rest of this case study aims to present what the company has done and the initial impact of GlaxoSmithKline's ethical approach in transforming the way pharma industry markets its prescription drugs.

In 2013, GlaxoSmithKline (GSK), a global pharma major, began a new, more ethical approach to customer interactions. A year later, in 2014, when its bribery scandal broke out in China, the company accelerated its ethical approach and took some radical steps, such as:

- decoupling sales force incentives from prescription volume
- redefining its interactions with physicians, including stopping all payments to doctors to speak on behalf of the company
- making greater investments in its people to significantly improve their knowledge, skills, and capabilities
- Increasing transparency measures across the board

**Putting Patients First**

Andrew Witty pledged to help patients with a disproportionate share of the world's disease burden. In February 2009, he initiated another transformational move for big pharma by cutting the prices of GSK's medicines in the least developed countries to no more than a quarter of their cost in the UK, and promised to reinvest twenty percent of the company's profits from those drugs to help those countries. Furthermore, he put 800 patents on potential drugs or parts of drugs into a pool so anyone can investigate them to find cures for neglected diseases such as malaria. Finally, Andrew Witty invited scientists to share GSK's labs for such research in Tres Cantos, Spain.

In an unprecedented move, Andrew Witty offered generic manufacturers in Uganda and other third-world countries royalty-free licensing of certain drugs to generic manufacturers if they applied to GSK for a license as long as they maintained the manufacturing quality.

When questioned about the impact that such a royalty-free approach on the company's balance sheet, Witty said: *This wasn't much in money terms. We bent over backward to ensure people knew, for example, that our profits in the LDCs (Least Developed Countries) were less than £5 million. I knew immediately people would think that must be billions. It's not. It's tiny numbers. It's the point of principle. Even £1 million is a lot of money in these markets. And if GSK does it, why can't every foreign company that makes profits in Uganda do the same? I don't just mean drug companies. Everybody.*

Whether Andrew Witty is leading GlaxoSmithKline into a revolutionary change— whether for idealistic reasons or to stem the bad publicity the company attracted in recent years you cannot ignore or overlook his concern regarding making medicines accessible to patients globally, including the LDCs. The sincerity of feelings is palpable when he says: *We can't just sit back and say we developed a drug and for some reason which we're not going to spend the time to try and understand, nobody in Africa can get it. I don't think that's an acceptable position to take. It's equally obvious we can't solve the problem because it is so big it's so complicated, but we can challenge ourselves to do more every day.*

In their 2013 rollout of a bold initiative to dramatically reshape how they interact with healthcare professionals, GlaxoSmithKline said: *In all our interactions with HCPs, we aim to be transparent about our work, operate with integrity, and always put the interests of patients first. We have made these changes because we believe the established operating method is outdated and ineffective for patients, HCPs, or the pharmaceutical industry*.

Further, Murray Stewart, GSK Chief Medical Officer, said: *These changes are significant, but we are committed to transforming our business model so that patients are at the heart of every decision we make*.

**GlaxoSmithKline's Ethical Model**

GSK made some foundational changes to:

- the way the company interacts with HCPs, how it incentivizes its sales representatives,
- how it supports medical education, and stopped payments to external HCPs for speaking on behalf of the company about its prescription medicines

**Why did GSK undertake this radical transformation?**

Because when someone visits a healthcare professional, they expect the advice and treatment recommendations they receive to be based on their medical needs and their HCP's understanding of their condition and the treatment options available. Any question of conflict of interest between a doctor and the companies that develop treatments could undermine the trust patients have in their treatment decisions. Also, if the information pharma companies provide to physicians is perceived as conflicting, this negative perception distracts from the treatment's core message and value proposition. It creates an unnecessary barrier to patient access. GlaxoSmithKline aimed to remove the opportunity for a conflict of interest and thus ensure that prescribers have access to adequate, objective, and current information that they can use to select the right treatments for their patients.

Has GSK changed its commercial objectives by shifting to its new *Ethical Model* of engaging HCPs? George Katzourakis, Senior Vice President at GSK, clarifies this point when he says: *Our commercial goal hasn't changed. It is simply a different way of going about it. The success of any pharma company depends on maximizing the number of patients that would benefit from its medicines. What is changing is how we communicate and engage with physicians, address their questions and concerns, and help identify patients who can use our medicines. At the same time, it addresses some reputational aspects and removes some conflicts of interest*.

**The Anatomy of GSK's New Ethical Model**

Every way you look at it, GlaxoSmithKline's new ethical model of engagement is pathbreaking. It is an unprecedented move that includes five major shifts from the past and current way of doing business.

1. **Decoupling sales force incentives from prescriptions:** While building and managing customer relationships, salespeople often face many ethical dilemmas. Therefore, GSK decoupled its sales force incentives from its sales targets and the number of prescriptions generated from HCPs as a key motivator for their sales representatives to reduce ethical dilemmas and deepen customer relationships.

   Under the new ethical model, GSK evaluates and rewards its representatives on three factors— selling competency, customer evaluations, and the overall operating profit of the pharmaceutical business.

A. ***Selling Competency*:** Reps are evaluated on scientific knowledge, proficiency, and understanding of the drug, technical skills, quality of service delivered to physicians, and business planning skills and execution. The placing of education as a key part of the incentive scheme increases reps' confidence level and sense of authority when communicating technical and medical information increases tremendously. The following questions indicate how GSK assesses its sales reps and sales managers in terms of their sales competency:

- Do reps have enough knowledge to represent the company's medicines and vaccines in front of their customers?
- Are they capable of delivering sales calls that bring real value?
- Can they differentiate their activities to customers where they can make the most difference?
- Is the scientific knowledge of managers enough to support their teams?
- Can they coach their sales teams to deliver sales calls of real value?
- Can they make the choices in their activity and resourcing to ensure that we see the customers where we can make a real difference?

B. ***Customer Evaluations*:** The company randomly collects customers' feedback to assess each sales rep's performance, capability to communicate the benefits of its products, and the quality of dialogue they maintain with the physician. This new approach allows sales reps to focus on delivering real value to their customers and fundamentally changes the attitude and approach to selling across the sales force. Further, the company also observes and rates reps on the quality of their communications with customers. With new methods like these to evaluate sales rep performance and engage them with the external medical community, internal medical and commercial teams have been allowed to demonstrate their skills and knowledge in new ways. In addition, by decoupling incentives with the sales and prescription targets, sales reps have become more intrinsically motivated to increase their customer understanding, their ability to ask questions that help identify the needs of the customer, and become a leading and valued information resource for physicians.

Victoria Williams, Vice President and Sales Director of GSK, France, says: *this approach drives a different behavior with physicians. Selling isn't a bad word if it's done right*!

2. **Redefining Interactions with Physicians:** In 2016, GlaxoSmithKline stopped paying physicians for speaking about the company's medicines and vaccines to other prescribers at medical conferences. Because when you make payments to prescribers, people link it to prescriptions and perceive that you are influencing the prescribing behavior of the physicians you pay.

3. **Greater Investment in People and Transparency Measures:** Denise Dewar, Head of Multichannel Marketing Excellence, European Union, Canada, and EMA (Europe, MiddleEast, and Africa) at GlaxoSmithKline, says that the company has invested heavily in up-skilling its medical, sales, and marketing teams and also brought in new talent that understands the new direction of the organization. Also, the company has invested in improved multichannel, particularly digital capabilities, to help strengthen its engagement capabilities with doctors. These new capabilities allow customers to share information to generate insights into what customers need.

In addition, the new digital capabilities allowed GSK to know more about drug profiles, safety, and efficacy parameters, and the significance of new treatments for patients, and the in-house medical team has been using webinars, peer-to-peer discussions, and social media platforms to speak to a large external network of physicians. Also, the company has launched 'Click to Chat' in some markets, allowing customers to instantly get in touch with a GSK doctor to ask a scientific query about a product or disease.

Another initiative GSK took to improve its transparency measures was creating a secure online portal, clinicalStudyDataRequest.com, in 2013 that allows researchers to request access to anonymized patient-level clinical trials and the number of clinical studies carried out by other pharma companies. In 2015, the biomedical research charity, the Wellcome Trust took over the management of the portal.

**GSK Tweaks its Ethical Approach on Physician Payments**

In June 2016, GSK disclosed 'transfers of value to HCPs across Europe in line with EFPIA (European Federation of Pharmaceutical Industries and Associations) requirements. Furthermore, the company disclosed in the true spirit of transparency, following a *no consent, no contract* approach.

GSK continued to pay physicians and HCPs for expert services and participation in clinical research endeavors basing the payments on fair market rates and subjecting them to rigorous control and transparency measures. Also, the company strives to sponsor conferences or talks held in more realistic locations with reasonable accommodation and hotel transfer services.

**Supporting Medical Education**

GSK has also changed the way it supports medical education. While the company still provides opportunities for medical education, they no longer play any role in selecting the prescribers who attend the conference or talks they fund. Thus, they have taken themselves out of the programs they fund.

Instead, an independent professional body organizes the educational programs without influence or participation from GSK, while the company continues to provide grant funding to reputable education providers. Murray Stewart nicely sums up the company's policy by saying that *we don't play if we pay*.

Another radical change that GSK implemented is that instead of paying external physicians to speak on their behalf or directly control medical education programs, the company has increased the number and expanded the role of its internal medical professionals.

GSK doctors provide educational support to external prescribers and speak about products and diseases on GSK's behalf—as company employees. Because the company wanted to take the prescriber out of the equation, it's a GSK employee giving talks now whose incentive isn't to prescribe. This increases the peer-to-peer discussion between the medical community and in-house physicians.

## Overcoming Challenges

The big change from transactional to transformational marketing in pharma was a huge challenge. The company has cut bridges to the past to get people moving on the journey.

George Katzourakis explains that they faced three major challenges:

1. *Changing culture:* The big change from transactional to transformational marketing in pharma was a huge challenge. The company has cut bridges to the past to get people moving on the journey.
2. *The second major challenge* was that the old infrastructure was not supporting the new model, so they had to build many capabilities, skills, and competencies— medical, digital, and marketing to integrate.
3. *Building internal belief and confidence* was another big challenge; one way to address it was to share some early wins to share working on getting people motivated and engaged with the new model.
4. *Employee Resistance*: Whenever there is a major organizational change, it is natural that some people who cannot adapt to change will leave the company, and salespeople at GSK are no exception. Some sales reps felt the changes were not for them and left the company. The company facilitated this by engaging its sales managers, communicating that it was okay to let go of those who wanted to leave. The company reassured their managers by saying people should choose to leave rather than try to fit them into something that will never be for them.

### Overcoming the Initial Pushback from Customers

Resistance to change that too such a big transformational change from transactional marketing practices is bound to happen, and it did in GSK's case resulting in a pushback initially from customers and employees. Those employees who could not cope with and adapt to changing environment left the company. The customers' initial resistance or pushback raised four key questions, which GlaxoSmithKline addressed promptly and appropriately. Consider these, for example:

1. **Are you saying we are not ethical?**

   Some doctors were offended when the change process started and felt they were being challenged on their ethics. However, the company explained that they recognize doctors make good decisions in the interests of their patients, and the company, too, puts patients themselves above everything else by supporting doctors in their endeavors.

2. **Am I not senior or experienced enough?**

   There is a popular perception in medical communities that a professor is more senior than a disease expert, who is more senior than a general practitioner. GlaxoSmithKline's use of internal GSK doctors as speakers questioned the entire medical profession hierarchy. So how did the company address this and overcome the perceptual hurdle? GSK understood that the only way to overcome this was to up-skill their sales reps and doctors to ensure they were genuine experts in products and disease areas.

3. **Is the quality of the medical information high enough?**

   Initially, quite a few doctors felt that they were not going to the GSK meetings because they did not think the speaker was not high enough quality, as Murray Stewart, CMO at GSK, expressed. However, over time and with word-of-mouth— and as GSK doctors and disease experts delivered more and more talks with strong scientific data — they realized that GSK speakers were highly credible not only about their products but also the general disease area, which reflected in increasing attendance at the GSK webinars and scientific meetings. In addition, the company practiced a feedback-gathering mechanism to continuously improve the scientific content, and delivery at their webinars and meetings. At the end of each webinar and meeting, GSK asked participants whether it was worthwhile and whether they would attend again.

4. **Is this a cost-cutting exercise?**

   Many doctors thought GSK was trying to cut costs by stopping payments to external HCP speakers. The company had to explain that while it had stopped direct payments to physicians for speaking arrangements, it had increased investment in medical education and grant funding to change this perception.

### Does GSK's New Ethical Approach Work?

While positive results from the new ethical approach were slow, qualitative evidence suggests improvement in the company's relations with HCPs. Consider these indicators, for example:

- Despite some problems initially, GSK's new ethical model improved its relationships with physicians. According to GSK, doctors became more open to seeing GSK reps and more willing to discuss and provide feedback on its medicines and vaccines and the diseases they treat. As a result, doctors increasingly felt that they were being heard and *understood* rather than *talked* to. In addition, the medical community is impressed with the knowledge of our sales force and the high standards and quality of dialogue.
- In Canada, where GSK stopped payments to HCPs in 2015, GSK speakers delivered the same number of medical product information sessions by June as external HCP speakers did during all of 2014, reflecting the strong scientific knowledge and communication skills of GSK speakers.

- Increased transparency increased collaboration between GSK and physicians. Furthermore, in Norway, it has boosted communication between GSK and physicians. In Denmark, the change to an ethical approach has provided the opportunity to develop more innovative engagements with physicians, such as a debate facilitated by an ethics professor. Also, in France, the French government and pharmaceutical authorities recognized GSK as a company doing something different. All these demonstrate that stopping payments was not about avoiding doctors but building trust by having science-led conversations.
- There is more positive feedback coming from across the globe. For example, a female pediatrician in Vietnam thanked GSK for their weekend webinars, for they made a big difference to a busy working mother.
- Even on the sales front, GSK did well in 2016 with appreciable growth in topline, bottom line, and new product sales.
- More importantly, the sense of pride among GSK's sales force increased significantly, as reflected in a 2015 employee survey where 81 percent said they were proud to work for GSK. How do you know that you are working the right way? GSK sales reps, after the implementation of the new ethical approach to customer engagement, understand this better than anyone else. They know that they are doing the right thing if they are okay with the way they work appearing in a newspaper.
- More than 90 percent of HCPs said that their interactions with GSK medics were helping them make more informed decisions and so benefiting patient care in a European Study of Physicians. The survey also revealed that 85 percent of respondents rated GSK meetings 4 out of 5 on a five-point scale.
- Furthermore, the number of HCPs attending face-to-face meetings with GSK medical speakers increased in 2016 from 4,000 to nearly 40,000.

## GSK's New Ethical Model: What Do Other Pharma Companies Think?

What do other pharma companies and key opinion leaders think about GSK's new ethical approach to physician engagement? Understandably, the opinions of other pharma companies and thought leaders were a mix of encouragement and skepticism. You can classify them into four areas:

1. **Why change the sales model entirely?**

   While agreeing to the main stance GSK took concerning unethical payment practices, such as bribes and kickbacks, should not be tolerated, John LaMattina, former president of R&D at Pfizer and now a Senior Partner at Pure Tech Health believes that the new model was not the right move for GSK. Here are his reasons:

- Every company should sell ethically, but that doesn't mean we must change all sales and marketing practices. Moreover, companies promote their drugs aggressively because they believe in them. Otherwise, they should not have gone to market.
- Regarding the soft measures and metrics in the GSK's new model to evaluate and reward their sales forces' performance, you cannot observe, monitor, and supervise salespeople easily, so basing compensation on results assures the company that its sales reps and managers are performing their duties.
- Furthermore, the behavior and proficiency that GSK wants its sales reps to espouse should be expected of any sales rep regardless of their company. Sales reps are expected to know about their products, field, and competitors, and closely work with the needs of physicians to get them to prescribe their products.
- When you don't drive the business as well as it should be, you don't have enough money to do good things like helping Third World nations and supporting R&D.

2. **Is GSK Simply Overcompensating for Previous Mistakes?**

   Some industry observers believe that GSK has been overcompensating for its recent corporate scandals, such as its $3 billion payment for unethical marketing practices and a bribing scandal in China. Some others question where is the necessity to change the sales model entirely, particularly when transparency and trust initiatives, such as the Open Payment Database that reports speaking payments made to prescribers, have now become widely adopted.

3. **What is the Risk to R&D Budgets?**

   LaMattina believes that if you don't have sales, your R&D budgets will stagnate, if not shrink. He says: *The rule of thumb is investing 15 percent of a company's top-line revenues in R&D activities. When sales drop, so does the budget for exploring and developing new compounds*. However, Christopher Bowe, Strategic Advisor and Executive Consultant at C. Bowe & Co, believes that cutting back on R&D expenses is not entirely disagreeable. The strategic imperatives for solving today's challenges for pharmaceutical companies are business-model-and-cost-based.

4. **Will the Industry Follow GSK's Lead?**

   Debra Whitman, Chief Public Policy Officer of the American Association of Retired Persons (AARP), says that the GSK approach and other similar business practices will be a standard across the industry.

While the American Association of Retired Persons (AARP) and many others hope that the GSK approach becomes the standard practice across the industry, there is little sign that the industry will follow GSK's lead. Many companies are watching how successful GSK's new engagement and sales model will be. Atkins says people need convincing that the GSK model is working before anybody takes it up. I don't know what the signals for that will be— maybe if GSK becomes a runaway success and has started to get brilliant ratings in the popularity contest, or if the retention of their sales reps became way better than others. Anyway, there must be other measures of success.

Some industry experts opine that moving towards the new model was the right decision. Andrew Witty, the Chief Executive of GSK and the man behind the GSK's new ethical model, demonstrated leadership and courage in taking the unpopular decision that will ultimately see the industry restore its sagging reputation within the wider community. Some predict the new model will become a trend—some willingly, others dragged along with the current.

GSK officials believe that all they need is time. Neil Barnes, Medical Head Global Respiratory Franchise at GSK, expressed this sentiment best when he said: *It is like smoking on airplanes. People will look back and say: Did we really used to do that?*

Murray Stewart at GSK believes that, ultimately, other companies will follow suit but that the pace of adoption will depend on its customers. Physicians will soon have a choice between attending a 20-minute meeting at a 5-star hotel with posh amenities and entertainment, and participating in a scientific talk at a small hotel with no food or entertainment. They will increasingly revisit their values and intentions by joining medical meetings and talks. Doctors will go where they acquire the information they need to better care for their patients. Over time, doctors will not want to participate in activities with a negative perception or ethical discomfort.

## Changing Landscape

While no other company has yet taken steps as large as GSK, some have made selective changes to their marketing and sales models. Consider these, for example:

- In July 2016, Novartis announced that it would evolve its business practices, including reforms to sponsorships for congress attendance and limitations on HCP speaker payments.

- Bristol-Myers Squibb (BMS) in China announced it would end speaker payments.
- By January 2018, member companies of MedTech, an alliance between the European Diagnostics and Manufacturers Association and the Board of the European Medical Technology Industry Association, wanted to phase out direct sponsorships for HCPs to attend conferences.

### A Step in the Right Direction

Finally, Christopher Bowe says there is much to learn from GSK's experiment. Further, he said: *In my view, it is a step in the right direction in that it acknowledges that the status quo cannot hold forever and something new is likely to be needed in the future. They took a big step— it could be ahead of its time. And anything ahead of its time will probably have a lot of growing pains and skeptics.*

### Key Lessons from GSK's Experience

What can we learn from GSK's experience with its new ethical approach to physician engagement and new sales and marketing model? Here are some important lessons:

1. *Meaningful Feedback Can Improve Sales Rep Performance*: You cannot replicate the objectivity that the traditional compensation models offer with qualitative assessment. For example, you cannot replace the objectivity of a sales ranking table. However, you can supplement the black-and-white metric system of financial goals with a sales performance feedback system that is more immediate and meaningful to the sales reps as it helps monitor and modify, if necessary, the behavior of sales reps' performance in the field.
2. *Transform Selling into a Meaningful Purpose:* Aggressive and unethical marketing practices have made selling a bad word in general. GSK's new ethical approach can transform selling into a meaningful purpose as it focuses on more transparent,

balanced sales visits with representatives being measured on bringing value to their customers.

3. *Embedding a New Sales Model Takes Time:* You cannot change the inner motivations of your sales teams overnight. However, many effective sales representatives enjoy their role and like to see the value they add to the doctor and patient and help accelerate the change.
4. *A Bold Model Requires A Bold Leader:* Huge organizational and cultural shifts need strong leadership. While the GSK's new ethical sales and marketing model involved everyone in the organization from the ground up, the strong and bold leadership of their CEO, Andrew Witty, galvanized the entire company's transformation by living and breathing patient focus and ethical behavior.
5. *Rewards Reinforce the Behavior Change:* You reinforce the behavior you reward is axiomatic. In the new ethical model, GSK still provides bonuses and rewards to sales reps, but with a difference. The difference is an ethical incentive plan. So what is an ethical incentive plan? The basis for incentives and bonuses shifts from a strong emphasis on sales and prescription targets to a mixture of sales targets and behavioral outcomes designed to optimize the performance of sales reps.

   According to Sanofi's Bacon, the balanced and ethical incentive plan should answer some basic questions, such as:

- Are you fulfilling the compliant requirement associated with your industry?
- Is the plan aligned with the brand strategy?
- Is it driving the right behaviors?
- Is it doing the right thing for our customers?

6. *You Need A New Breed of Sales Reps:* A new ethical sales and marketing model essentially changes the role of future sales reps. You need people with different skills, capabilities, and attitudes.

Matthew Smith, Senior Manager of LifeSciences Advisory at Ernst & Young, says that *staying still or the same was not an option. While the jury is still out on GSK, it is the right move. It's a top-down change, which shows commitment. Even if it's not an overnight success, they will likely attract talent and build trust that has a lasting impact down the road*.

Finally, Christopher Bowe hits the nail on its head when he says: *GSK is looking at a long-term horizon. By changing the skills and aligning a sales force with health goals instead of pill volume, a company might create a new and sorely needed value for their customers by having a public health army out there in the field who are helping to drive goals beneficial to society*.

(Source: Adapted from an eyeforpharma article—GSK's new ethical approach: Is it delivering? A review, analysis, and key lessons— with contributions from Stewart Adkins, Marc Bacon, Christopher Bowe, John LaMattina, George Katzourakis, Matthew Smith, Murray Stewart, Jack Whelan, and Debra Whitman, Eyeforpharma White Paper, and an article from the Guardian— Sarah Boseley, Andrew Witty: Drug Firm Boss Out to Change His Industry, The Guardian, August 12, 2009)

There is a popular misconception that transformation of any nature, be it overall marketing approach or digital way of working requires deep pockets and therefore small and medium size pharmaceutical companies cannot afford it. The question is— does size really matter?

## Does Size Matter?

How important is the size of a firm for moving from transactional to transformational marketing in the pharmaceutical industry? Does size matter?

Size does not necessarily matter in shifting from transactional to transformational marketing in the pharmaceutical industry. Both small and large pharmaceutical companies can implement transformational strategies.

# Transactional to Transformational Marketing: Small Pharma

## Changing the Mindset

Changing the mindset is crucial for a small pharmaceutical company to move from transactional to transformational marketing. While deep pockets can help invest in the necessary technologies and resources, true transformational marketing requires a shift in thinking and perspective. Therefore creating a change vision is a first must. Here are four ways a change mindset is important to a small pharmaceutical company.

1. A transformational approach requires a shift from focusing solely on sales transactions to putting the patient at the center of everything you do. In other words, a patient-centric approach. This requires a mindset change from viewing patients as customers to viewing them as individuals with unique needs and goals.
2. Moving from a transactional to a transformational marketing approach in the pharmaceutical industry can be a significant shift. Still, connecting with patients and healthcare professionals is critical as it can lead to greater success. However, it requires an understanding of the difference between transactional and transformational marketing. Transactional marketing focuses on short-term sales, while transformational marketing focuses on building long-term relationships. This requires a mindset change from viewing each interaction as a one-time transaction to viewing it as an opportunity to build trust and loyalty.
3. Transactional marketing focuses on promoting products, while transformational marketing focuses on positively impacting patients' lives. This *means to an end, and* viewing it as a way to make a difference in the lives of patients and healthcare professionals is what makes transactional approach transformational.
4. A small pharmaceutical company may have limited resources but can still be innovative and adaptive. This requires a mindset

change from viewing limitations as obstacles to viewing them as opportunities to think differently and create new solutions.

### What it Takes For Small Pharma

With the mindset change in place, a small pharmaceutical company can invest in many technologies to move from transactional to transformational marketing in a phased manner. Here are a few technologies that could be considered.

1. *CRM (Customer Relationship Management) Software*: This technology can help a small pharmaceutical company manage customer interactions and data, track customer interactions across multiple channels, and create targeted marketing campaigns. Investment level: $1,000 to $5,000 per year.
2. *Social Media Management Platform*: A small pharmaceutical company can use social media platforms to connect with patients and healthcare professionals. A social media management platform can help automate and streamline these interactions. Investment needed: $100 to $5,000 per month.
3. *Email Marketing Platform*: Email marketing can be an effective way to reach patients and healthcare professionals, and an email marketing platform can help to automate and track these interactions. Investment level: $10 to $500 per month.
4. *Digital Analytics and Tracking Tools*: Digital analytics and tracking tools can help a small pharmaceutical company track the performance of its digital marketing campaigns and make data-driven decisions. Investment level: $1,000 to $5,000 per year.
5. *Virtual and Augmented Reality Technology*: This technology can create immersive and interactive experiences for patients, healthcare professionals, and other stakeholders. Investment level: $5,000 to $50,000 per project.
6. *Artificial Intelligence (AI)-Powered Chatbots*: AI-powered chatbots can provide 24/7 customer support, answer frequently asked questions, and assist with appointment scheduling. Investment level: $5,000 to $20,000 per project.

A word about the investment levels. Please note that these are rough investment levels, and the actual costs may vary depending on the specific technology, the size of the company, and its goals. A small company needs to consider its budget and also keep in mind that the investment in technology is just one aspect of the transformational marketing strategy and should be aligned with the overall goals and objectives of the company.

For example, a pharmaceutical company spends 15 to 30 percent of its sales on transactional marketing inputs, such as paying physicians on whatever pretext, expensive gifts, luxury holidays, and kickbacks. Well, that is all promotional expenditure for the year, which is written off. However, if the same company invests a major portion of that amount on transformational marketing inputs, such as technology, training, and acquiring the company's talent, patient-centric activities, product development, and addressing the patients' pain points and unmet needs, that would be an investment that accumulates value each year. And before long, the company would be tech-savvy, *winning with the new rules of engaging physicians, patients, and other stakeholders*!

**Small Pharma Also Can Transform**

Here are six steps for a small pharma company to move from transactional to transformational marketing:

1. *Start by assessing the current marketing approach*: Look at the current marketing strategies and determine which are transactional and which have the potential to be transformational. Identify areas where changes can be made to shift towards a more transformational approach.
2. *Develop a patient-centric approach*: Transformational marketing puts the patient at the center of everything you do. This means understanding their needs, wants, and goals and developing strategies aligning with them.
3. *Focus on building Trust*: Trust is a critical component of any relationship and is particularly important in healthcare. Building trust with patients and healthcare professionals is essential for a successful transformational marketing approach.

4. *Leverage digital tools and technologies*: Digital tools and technologies can help a company connect with its patients and healthcare professionals in new ways. Use social media, Email, and other digital channels to build relationships and provide valuable information.
5. *Measure and track your progress*: As you implement your transformational marketing, measuring and tracking your progress is important. This will help you to identify areas that are working well, and areas that need improvement.
6. *Partner with influencers and other healthcare organizations*: A small pharmaceutical company may have limited resources and may not have the same reach as a larger company. Partnering with influencers, healthcare providers, and other organizations can help you reach a broader audience and build trust with your target market.

Thus, a small pharmaceutical company can move away from its transactional marketing to transformational marketing by being innovative and adaptive with the resources they have at its disposal. They can invest in many technologies in a phased manner and be a tech-savvy company, building long-term relationships with all its stakeholders based on trust and positively impacting patients' lives.

The following case study of Curatio, a small pharmaceutical company operating in the fiercely competitive and predominately branded-generics market, such as India shows that pharmaceutical companies regardless their size, small, medium or large can follow a transformational marketing strategy.

CASE

23

# Curatio's Foray into Digital Marketing!

A small yet powerful company with a sharp focus on the dermatological segment in general and cosmetic dermatology in particular of the Indian pharmaceutical industry has been using digital tools very effectively in patient education. Dermatology accounts for about 7 percent of the total pharmaceutical market in India.

Curatio started about twelve years ago and has made rapid strides in the last three to four years. The company recorded ₹ 224 crores in annual sales in fiscal 2022. What is commendable about Curatio is that the company decided to adopt a transformational approach to marketing its products and services rather than transactional marketing, which most of the players in the specialty segments follow in the Indian pharmaceutical market. The smaller companies, in particular, rely heavily on gratification based or *quid pro quo* marketing methods. Their marketing plans include substantial discounts, expensive gifts, luxury travel, holiday junkets, and even cash for generating prescriptions. Curatio thought differently, took a brave decision, and instead invested in technology and started using certain digital tools to reinforce its marketing.

The company realized that physicians and patients increasingly use many digital tools daily. More and more patients are turning to the internet to search for more information on their disease conditions and the treatment options available and even finding out which physicians they should consult.

Physicians are becoming busier than ever and find no time to spend with medical representatives. The detailing times have been reduced to two minutes or less per visit. The time that a physician spends with his patients too has been shrinking. Patients have many questions, and physicians cannot answer to their patient's satisfaction. Here is a need that a pharmaceutical company can ideally fill. Curatio identified this gap and formulated a patient-centric strategy creatively to meet this unmet need. The company planned to use digital tools to realize this opportunity and fill the gap.

The company has planned to use four of the most popular digital tools to educate patients on behalf of their physicians. One is the use of customer care through telephone services. The difference here is that the telephone service of Curatio is not an automated call-center service that is annoying by directing the caller to press many keys until he reaches the company's customer service executive. Instead, the call directly goes to a responsible product executive, who immediately responds to the query. In addition, the company has given the customer service contact number on the packs of the respective products. With this direct telephone service, the company achieved two significant wins — one is customer satisfaction, and the other is priceless insights into customers' needs and wants!

The other digital tool that Curatio has used is the ubiquitous YouTube for educating patients on how to use their products once they are prescribed. The company used this elegant engagement method on behalf of the physician. It did not bypass the physician. The Physicians, too, are happy that the company has been assisting them in delivering useful patient education information leading to better compliance.

Also, Curatio has used an e-Detailing platform to deliver highly effective communication about its products and services to HCPs when they want it. e-Detailing is available round the clock, 365 days a year.

Another important digital channel that the company has used is blogs. Curatio practiced influencer marketing by roping in top blog writers and Instagram influencers to write on relevant topics related to their products. These blog articles targeted at the consumers helped the company build greater awareness and a favorable image.

These digital tools have significantly reinforced the company's overall marketing effort, showing many companies that they can grow rapidly through transformational marketing.

The success of Curatio is indeed an eye-opener to all those companies heavily dependent on transactional marketing practices. Transactional marketing may appear to give immediate returns, but they are indeed in the short term. Transactional marketing is not sustainable as the relationships are very transient. A slightly higher discount, a marginally better offer, could whisk your customers in a jiffy. Moreover, the impending legislation may declare transactional marketing practices illegal or criminal. Transformational marketing will no longer be an option. Sooner or later, it will become mandatory! Curatio shows the way to transform.

## Transformational Marketing in Pharma: Final Thoughts

While implementing transformational marketing practices in your organization, there is one thing that matters most and need serious contemplation, reflection, and consideration. It is:

## Making Selling and Marketing Respectable Again!

Transformational marketing in pharma is all about making marketing respectable again because it has been a highly respected industry until a few decades ago.

While it is difficult to pinpoint when and why the industry's reputation and respectability began sliding, consider these plausible reasons and timelines.

One is the Waxman-Hatch Act of 1984, which facilitated the approval of generic drugs in the United States. The other is the increasing genericization of the pharmaceutical market has likely contributed to the increased pressure on pharmaceutical companies to meet quarterly earnings expectations, leading to unethical marketing practices.

The rise of generic drugs has intensified competition in the pharmaceutical industry, increasing pressure on companies to maintain profitability. Unfortunately, this has led some companies to engage in aggressive, sometimes illegal, and unethical marketing practices, such as off-label promotion and kickbacks to healthcare providers to boost sales and meet quarterly earnings expectations. Thus, the pharmaceutical industry has entered into what may be called a transactional marketing era in pharma.

How can pharma get back to the transformational mode of marketing that it has been in the past?

Pharma has to make its marketing respectable again by following a few simple processes that it was used to in the past and adapting and absorbing new and emerging technologies that are revolutionizing the way work is done today.

1. *Stop all unethical and illegal marketing practices*, such as off-label promotion and payments to physicians for prescribing

your products in cash or kind. Start focusing on science-based promotion and communicating transparently.

2. *Institute ethical behavior in all your team members*. You can achieve this by clearly defining ethical behaviors and reinforcing them by rewarding them. Remember the motivational axiom—you reinforce the behavior you reward! Ensure your incentive plans balance qualitative and quantitative KPIs (Key Performance Indicators).
3. *Transformational marketing needs a new breed of sales force* proficient in the science behind the products and diseases, and tech-savvy to orchestrate the multi and omnichannel engagement strategies effectively for building enduring relationships. Also, you must invest continuously in training and development to keep them ahead of the competition.
4. *Payments to physicians*: If you have to pay physicians for the services they render to your company, such as speaking at symposia, clinical trials, and consulting on new product development, ensure you make these on a market-value basis and pay for their services, and not as kickbacks. Also, when you sponsor any symposia for physicians, ensure you only pay and not play. This means not taking part in deciding the attendees or the topics or speakers for the symposia.
5. *Put patients at the center of everything you do*. Patient focus and patient-centricity are of paramount importance to pharmaceutical marketing.
6. *Reward the behavior you want to reinforce.* The principle of positive reinforcement emphasizes the need to reward the behavior you want to change. For example, consider the question— Why do people do what they do? The answer is—People do what they do because of what happens to them when they do what they do. So if you make the consequences for the desired behavior (Ethical Selling) positive (by rewarding and recognizing), they will repeat their behavior (Ethical Selling). Therefore, implement an incentive plan that rewards Ethical Selling that gives at least equal importance to the ethical behavior on selling (customer satisfaction, scientific selling,

selling competencies, and others), prescriptions, and sales targets.

These are the six most important points to remember and act upon in making your marketing efforts respectable again. They are not as difficult as they seem. Moreover, that is what industry leaders have been doing in the past, precisely seventy-three years before. All we have to do is remember constantly what George W. Merck, the legendary Founder and Managing Director at Merck, said in 1950:

*We try to remember that medicine is for the patient. We try never to forget that medicine is for the people. It is not for the profits. The profits follow, and if we remember that, they have never failed to appear. The better we remembered it, the larger they had been.*

CHAPTER

# 8

# Transformational Marketing: The Winner's Checklist

*Under conditions of complexity, not only are checklists a help, they are required for success.*

— Atul Gawande

# Transformational Marketing: The Winner's Checklist

You can never overemphasize the importance of a checklist as a vital tool for improving productivity. In the book, The Checklist Manifesto: How to Get Things Right, the author, Dr. Atul Gawande writes that: *Good checklists are precise, to the point, and easy to use even in the most difficult situations. They do not spell everything—a checklist cannot fly a plane. Instead, they provide reminders of only the most critical and necessary steps—those that even the highly skilled professional using them could miss. Good checklists are, above all, practical.*

### What is a Checklist?

The Collins Dictionary defines a checklist as a list of all the things you need to do, the information you want to find out, or things you need to take somewhere, which you make to ensure that you do not forget anything.

### The Evolution of Checklists

Boeing, the leading Aircraft manufacturer, created the first-ever checklist following a deadly crash of their new model 299, known as the Boeing B-17 Flying Fortress, on October 30, 1935. However, an investigation found the airplane to be in perfect condition. No mechanical failure could be found that would have resulted in the deadly crash. After a deeper investigation and eyewitness reports, including the surviving co-pilot, it was determined that the flight crew had forgotten to release the flight control gust locks, which had caused the plane to nosedive into the ground immediately after takeoff. How did it happen? It was further determined that the airplane was too complex for man to fly. Far too many tasks were required to safely take off a modern aircraft, making it extremely difficult to remember by humans.

Boeing introduced the *Checklist* after the crash of the Model 299 on October 1935 to be used by all pilots in the Boeing Fleet. After introducing the checklist, Boeing could sell the Model 299 to the US Army Air Corps, which flew for years without incident.

The use of checklists in aviation quickly became widespread, with pilots using them to verify that all necessary steps were completed before takeoff and landing, such as checking the fuel levels, engine controls, and navigation systems. The checklists increased safety and reduced the risk of human error, leading to the widespread adoption of checklists in other industries, such as medicine construction, finance, and marketing.

### Checklists in Pharma Marketing

Although simple tools, checklists can profoundly impact the effectiveness of pharmaceutical marketing. In a highly regulated industry, such as pharmaceuticals, where even minor mistakes can have serious consequences, checklists provide a systematic way to ensure all the necessary steps are taken, and all the required information is gathered and analyzed before making decisions and executing plans. Therefore, improving efficiency is one of the primary benefits of Checklists in pharmaceutical marketing. By having a clear and comprehensive list of tasks to be completed, marketers can save time by focusing on essential tasks. Furthermore, checklists help pharma marketers avoid unnecessary tasks or forgetting important steps. This helps ensure that marketing tasks are completed promptly, reducing the risk of missing deadlines and losing opportunities.

In addition, checklists help enhance accuracy. Checklists provide a systematic way to ensure all necessary information is gathered and analyzed, reducing the risk of errors or oversights. Thus, using checklists, marketers can ensure that all relevant information is considered, reducing the risk of making decisions based on incomplete or inaccurate information.

Here are seven important reasons why checklists are essential in pharmaceutical marketing.

1. *Consistency:* Checklists ensure that all marketing efforts are consistent and follow established best practices and guidelines.
2. *Accountability:* Checklists help teams and individuals stay accountable for their tasks and responsibilities by providing a clear roadmap for what needs to be done and when.

3. *Standardization:* Checklists standardize marketing processes and procedures, making it easier for teams to collaborate and work together effectively.
4. *Efficiency:* Checklists help streamline marketing efforts and reduce time and effort required to complete tasks and achieve goals.
5. *Improved Quality:* Checklists help to improve the quality of marketing efforts by ensuring that all necessary steps are taken and that nothing is overlooked.
6. *Alignment with Regulations:* Checklists help ensure compliance with industry regulations and standards, which is critical in the pharmaceutical industry.
7. *Better Decision-Making:* Checklists provide a structured framework for decision-making, helping teams to make informed decisions based on best practices and data-driven insights.
8. *Valuable Tool for Training and Development:* Checklists can also be a valuable tool for training and development in the pharmaceutical industry. New employees can quickly understand the requirements and their role because checklists provide a clear and comprehensive list of tasks and procedures, helping reduce new hires' learning curve and improving overall productivity.

Checklists are an essential tool for managing complex marketing efforts and improving outcomes in the highly regulated pharmaceutical industry. They help pharmaceutical marketing teams achieve their goals more efficiently and effectively and deliver better results for their organizations and patients.

These are the eight most important benefits that checklists offer pharmaceutical marketers to help improve efficiency, effectiveness, accuracy, and overall productivity in their marketing efforts.

### Two Levels

Pharmaceutical companies that want to move from transactional to transformational marketing should implement a company-wide transformation program at two levels— ethical behavior and digital transformation. This is because changing healthcare environment demands it. So what is the major change in the healthcare environment? Increasing healthcare consumerism is top of the list of changes.

### Consumerization of Healthcare

The consumerization of healthcare refers to the shift towards a more patient-centric model, where individuals have greater control over their health and healthcare decisions. Several factors, such as technological advances, healthcare delivery changes, and shifting consumer expectations, drive this change.

One of the most significant drivers of the consumerization of healthcare is the proliferation of digital health technologies. From wearables and mobile health apps to telemedicine and virtual consultations, these technologies empower patients to take a more active role in managing their health. For example, individuals can track their vital signs, monitor chronic conditions, and access medical information and advice through smartphones. This has enabled patients to make more informed decisions about their health and seek care when and where they need it.

Another key driver of the consumerization of healthcare is the shift towards value-based care. With increasing focus on reducing costs and improving outcomes, healthcare providers are increasingly looking for ways to engage patients in their own care and educate them about the best ways to manage their health. This includes providing patients with tools and resources to help make more informed decisions, such as cost and quality transparency, and empowering them to take a more active role in their care through self-management programs. Consumers also demand more convenient, accessible, and personalized healthcare. The rise of retail clinics, telemedicine, and other alternative care models are meeting these demands by giving patients more options for where and when they receive care. Patients are also increasingly seeking healthcare providers who can deliver personalized, tailored care based on their needs and preferences.

When you carefully examine, you find that the *Major Drivers* of change are in areas like:

- Information technology
- Global perspective for global markets
- Ethical corporate behavior
- Responsibility to society at large, not just shareholders
- Total customer solutions and customer-centricity

The rate of change in the global pharmaceutical industry is so high that incremental improvements will not work, let alone maximize your potential. Gone are the days when you could gain a competitive advantage by developing a slightly better way of doing things. The pace of change is so rapid that it is almost like a gale-force wind. You will probably go back if you aim to take a small step. You need to take a big step into the teeth of the gale merely to hold your position. You must take more than one big step at a time to move forward. The change has to be revolutionary, not evolutionary.

Consider what Gary Hamel said while discussing strategy as a revolution: *Let us admit it. Corporations around the world are reaching the limits of incrementalism. Squeezing another penny out of costs, getting a product to market a few weeks earlier, responding to customers' inquiries a little bit faster, ratcheting quality up by one or more notch, and capturing another point of market share—these are the obsessions of managers of today. But pursuing incremental improvements while rivals reinvent the industry is like fiddling when Rome burns.*

While the usual is no longer usual, a business-as-usual approach will fail. Moreover, the rate of change offers opportunities for competition to exploit if you fail to do so. John O'Keefe, Group Vice President, Proctor & Gamble, in his highly stimulating and thought-provoking book *Business Beyond the Box: Applying Your Mind for*

*Breakthrough Results*, says that: *Too many organizations use the past as the sofa. The past should be used as a springboard and not as a sofa*. The critical question is how your organization sees the past—as a sofa or a springboard.

The Winners' Checklist, in our context, is about asking some of the essential questions that should signal to a company how the change is taking place and how much 'it' needs to change. To win in the changing healthcare marketplace, pharmaceutical companies need to monitor the progress constantly and continuously at two levels of their transformation program— Ethical behavior and Digital transformation.

# Monitoring Ethical Behavior in Pharma: A Checklist

1. **Compliance with laws and regulations:** Are our marketing materials in compliance with local and international regulations and laws, such as FDA guidelines, data privacy regulations, and anti-bribery regulations? Are we following the ethical codes, such as IFPMA, PhRMA, EFPIA, UCPMP depending upon our affiliation?
2. **Accurate representation of product information:** Is all our product information accurate, up-to-date, and not misleading? Are we desisting from promoting our product for off-label and unapproved conditions?
3. **Fair Comparisons:** Are we avoiding unfair comparisons with competitor products or making false claims about the product's efficacy?
4. **Informed Consent:** Are we ensuring that patients are fully informed about the risks and benefits of using the product and have given consent to participate in any marketing activities?
5. **Transparency in interactions with healthcare professionals**: Are we disclosing any financial relationships or incentives offered to healthcare professionals and ensuring that such interactions comply with ethical guidelines? Are we avoiding offering gifts or incentives to HCPs that could influence their prescribing behavior?
6. **Responsibility in marketing to vulnerable populations**: Are we ensuring our marketing activities are conducted with care and sensitivity when targeting vulnerable populations, such as children and elderly patients?
7. **Proper use of influencer marketing:** Are we ensuring that influencer marketing is conducted transparently and ethically and that influencers disclose any financial relationship or incentives?

8. **Monitoring and Reporting**: Have we established a mechanism for employees to report unethical behavior and conduct regular audits to meet ethical standards?
9. **Respect for Patient Privacy**: Are we respecting patient privacy and protecting it by not sharing it without proper consent?
10. **Employee Training**: Are we providing regular training to employees on ethical considerations in pharmaceutical marketing and the consequences of violating ethical standards?

## Digital Transformation: A Checklist

1. Do we have a well-defined *Mission* (not just a statement) that all team members understand? How do our senior management team members see themselves as – revolutionaries or evolutionaries in our industry? Do they contend with the status quo? Or do they want to rewrite the rules for the industry? What is our leadership quotient? What is their foresight about our industry?
2. Do we have a clear *Vision* and a detailed blueprint for action on how we will achieve what we want? How detailed are our action plans? Do they indicate responsibility centers and timelines for each key objective?
3. Do all our employees share organizational optimism and aspirations? Do they have a clear sense of urgency and a sense of ownership of our organizational goals and objectives?
4. Do we have a clear understanding of our target audience and their needs?
5. Do we have a clear brand strategy and messaging that is based on science?
6. What are the appropriate channels for reaching the target audience and delivering the content?
7. Do we have a data-driven approach to measure and optimize our marketing performance?
8. Does our current customer experience match the desired customer experience? What is the gap, and do we have a time-bound specific action plan to bridge it?
9. Can our existing digital estate deliver to customers their actual needs?How have our competitors and peers adapted to new conditions shaped by customers' digital expectations?
10. Do we continuously and constantly upgrade our technologies and train our teams on those regularly to stay ahead of the competition?
11. How are competitors or alternatives in the marketplace meeting customers' needs better than ours?

12. What areas of our business present the greatest barriers to digital transformation?
13. What areas of our Business require the most transformation?
14. What are the critical barriers to achieving our digital transformation vision?
15. What skills, capabilities, and resources are we missing to undertake transformation efforts?
16. How do we stack up against the competition? Are our competitors' product and service offerings similar to ours? What is our competitive advantage?
17. Can the goals be measured against the five-point evaluation criteria— customer value, competitive advantage, business impact, technical feasibility, and scalability?
18. Do we have an effective monitoring mechanism to track and analyze key metrics to measure marketing performance, and monitor the competition and industry trends to stay ahead of the curve?
19. Do we collect and analyze customer feedback to inform future marketing conditions?
20. What is the cost and impact on the existing systems?

Please note that these two checklists for monitoring the ethical behavior and digital transformation efforts are intended to serve as a general guide. Pharma marketers should develop detailed checklists to monitor every key aspect of their transformational effort covering their marketing, technology and ethical initiatives.

### Winning Leadership for an Era of Change

Lawrence A. Bossidy, an American author and former CEO of Allied Signal (later Honeywell), suggested four action steps for making a company a winner in an era of change through effective leadership many years ago, which are relevant even today as they are axiomatic.

1. First, determine the task by honestly assessing your stand with customers, employees, and competitors.

2. Two, articulate a vision of where you want to be and the values for your behavior. Set aggressive goals and keep everyone focused on achieving them.
3. Three, fill your company with people bursting with energy and creativity— people with diverse talents who can work together in a team setting. Give people constant, candid feedback, and let them know that the organization can fulfill their highest career aspirations.
4. And finally, create a culture obsessed with customers, and build your organization and incentive systems around the customer's interests.

# Epilogue

## Remember Rip Van Winkle?

Most of the people in this world are asleep in their minds.

*— Seth Adam Smith, Blogger and Author of Rip Van Winkle and the Pumpkin Lantern*

# Epilogue
# Remember Rip Van Winkle?

It is likely that most companies in branded-generics markets, such as India, think that disruptive technologies will not impact their markets much. Therefore, adopting the expensive new and emerging technologies that big pharma embraces is unnecessary. They may even believe they can continue their business as usual with the transactional, *quid pro quo* practices they are comfortable with. All those marketers and entrepreneurs who think they can survive and thrive with transactional marketing practices need to remember Washington Irving's most famous piece of writing—the story of *Rip Van Winkle*. It offers invaluable lessons.

For those of you who do not remember the story of Rip Van Winkle or have not read it, here is the story in brief.

## Rip Van Winkle

Rip Van Winkle, a kind man, and a henpecked husband lives in a village near the Catskill Mountains in New York. He is dutiful, always quick to help his friends and neighbors, and well-liked. Besides his termagant wife, he has children, including a son named Rip, who resembles his father strongly.

Rip Van Winkle also has a dog, Wolf. Rip never likes working on his farm but loves spending time with other men in the village at a local pub named after King George the Third of Great Britain, gossiping and discussing other topics.

One day, Rip Van Winkle walks up the Catskill Mountains with his dog Wolf. As he is about to descend, he hears someone shouting his name. A strange, short man with a grey beard appears, wearing antique Dutch clothes. He beckons Rip to follow him, and they arrive at the woodland amphitheater where strange people are playing ninepins. They are also dressed in old clothes. The man who has led Rip here has a keg of alcoholic drink, which he shares with these figures.

Rip Van Winkle tries the drink and takes such a liking to it that he drinks too much of it and sinks into a deep sleep. When he wakes up, all of these strange figures have gone, including the man with the keg of liquor. Rip's dog has also gone. The gun he took with him up the mountain has gone, and a rusted gun is next to him instead.

As he walks home, Rip Van Winkle realizes his beard has grown a foot long. When he arrives back in his village, he meets people he doesn't know and who don't know him. All of the shops and houses look different. When he goes into his home, it's to find that it's rundown and deserted. Going out into the street, he finds that the pub he used to meet with his friends outside has changed King George the Third to General Washington.

Rip Van Winkle speaks with the villagers and asks if they know of two of his oldest friends, whom he names. They tell him that those two friends have died. Next, Rip Van Winkle asks them if anyone knows a man named Rip Van Winkle. They point out to a man who looks just like Rip— his son, now grown up and resembling his father.

Rip Van Winkle's daughter, also grown up, appears with a baby. Rip Van Winkle asks her who her father is. She replies that his name was Rip Van Winkle, but that he disappeared twenty years ago after he went for a walk in the mountains. They feared he had been captured by Native Americans or had shot himself. It turns out that Rip Van Winkle thought he had slept for one night, but he had, in fact, been asleep for twenty years.

Rip Van Winkle asks his daughter what happened to her mother (Rip's wife) and is relieved to know that she has died.

Later, Rip Van Winkle tells his story of the night he went to the mountains and settled down with his family, watching his grandchild grow while his son tended to the farm.

Rip Van Winkle eventually reacquaints with his remaining friends in the village. He becomes revered as a village elder and patriarch who remembers what the village was like before the American Revolutionary War.

Thus, Rip Van Winkle has become a byword for the idea of falling asleep and asking to find the familiar world around us has changed beyond recognition.

(Source: Dr. Oliver Tearle's article, A Summary, and Analysis of Washington Irving's 'Rip Van Winkle' published in Interesting Literature Blog. Dr. Oliver Tearle is a literary critic and lecturer in English at Loughborough University in the UK.)

**What is the Moral of the Story?**

The story of Rip Van Winkle tells us that you cannot turn a blind eye to change and ignore it. You cannot expect the status quo to continue, and it will be business as usual. The story also shows us that people will pay dearly when they try to avoid change. They better be active participants in the changing environment.

Therefore, pharma marketers in branded-generics markets will be better if they heed this message and embrace the new and emerging technologies disrupting their business environment. Or else, they will go into a deep sleep like Rip Van Winkle did, only to find out that the business world around them has changed so dramatically and beyond recognition and will not have any clue to participate in it. They will likely end up as *Pharma Elders* to narrate how the pharmaceutical industry was before it was disrupted to those curious people around them.

The message is loud and clear. Embrace the change or become a part of history in the pharmaceutical industry.

# References

1. Marcia Angell, M.D, The Truth About the Drug Companies: How They Deceive Us and What To Do About It, Random House, New York, 2004
2. Joel Lexchin, MD, Doctors in Denial: Why Big Pharma and the Canadian Medical Profession are Too Close for Comfort, Kindle Edition, James Lorimer & Company Ltd, Toronto, 2017
3. Melody Petersen, Our Daily Meds: How the Pharmaceutical Companies Transformed Themselves into Slick Marketing Machines and Hooked the Nation on Prescription Drugs, Picador, First Edition, 2009
4. Katherine Eban, Bottle of Lies: The Inside Story of the Generic Drug Boom, Kindle Edition, Ecco, Reprint Edition, 2020
5. Stephen Sheller, Sidney Kirkpatrick, Pharmageddon: A Nation Betrayed: A National Trial Lawyer Reveals an Industry Spinning out of Control, Kindle Edition, Cape Cedar Media, 2016
6. Jerome P. Kassirer, MD, On The Take: How Medicine's Complicity With Big Business Can Endanger Your Health, Oxford University Press, 2005
7. John L. LaMattina, Devalued and Distrusted: Can the Pharmaceutical Industry Restore its Broken Image, 1st Edition, Kindle Edition, Wiley, 2012
8. Robert Yoho, MD, Butchered By "Healthcare": How Doctors and Corporations Try to Kill you For The Money and How To Survive Anyway, Amazon Kindle Edition, 2021
9. Gerald Posner, Pharma: Greed, Lies and The Poisoning of America, Simon & Schuster,
10. Christopher Lane, Shyness: How Normal Behavior Became a Sickness, 1st Edition, Kindle Edition, Yale University Press, 2007
11. Alison Bass, Side Effects: A Prosecutor, a Whistleblower, and a Bestselling Antidepressant on Trial, Kindle Edition, Algonquin Books, 2008
12. Dr. Arun Gadre, and Dr. Abhay Shukla, Dissenting Diagnosis: Voices of Conscience from the Medical Profession, Random House India, 2016

13. Edited by Samiran Nundy, Keshav Desiraju, Sanjay Nagral, Healers or Predators? Healthcare Corruption in India, Oxford University Press, 2018
14. Shailendra Tripathi, CRM in Pharmaceutical and Healthcare Marketing, Clever Fox Publishing, Manipal Technologies Limited, 2020
15. John Abramson, Overdosed America: The Broken Promise of American Medicine, Harper Collins Publishers Inc, 2008
16. Atul Gawande, The Checklist Manifesto: How to Get Things Right, Profile Books, London, 2010
17. Linda Banks, Applying Closed Loop Pharma Marketing in Pharma, pharmaphorum.com, September 9, 2014, https://pharmaphorum.com/views-and-analysis/applying-closed-loop-marketing-in-pharma/
18. Ramon Chen, Data Driving Metamorphosis of Pharma Marketing, pharmaphorum.com October 26, 2016, https://pharmaphorum.com/views-and-analysis/data-driving-metamorphosis-of-pharma-marketing/
19. Robin Robinson, Marketing Using Data Analytics, pharmavoice.com, March 1, 2018, https://www.pharmavoice.com/news/2018-03-analytics/612645/
20. Pradeep Khedkar and Saby Mitra, Boosting Pharmaceutical Sales and Marketing with Artificial intelligence, ZS Associates, https://www.zs.com/content/dam/pdfs/Boosting_Pharma_Sales_With_Marketing_AI.pdf
21. Denis G. Arnold, Louis H. Amato, Jennifer L. Troyer, Oscar Stewart, Innovation and Misconduct in the Pharmaceutical Industry, ELSEVIER Science Direct, Journal of Business Research, https://doi.org/10.1016/j.jbusres.2022.02.026
22. Phanish Chandra, Pharma Marketing and Ethics— Will the Twain Ever Meet? LinkedIn Blog, February 15, 2017, https://www.linkedin.com/pulse/pharma-marketing-ethics-twain-ever-meet-phanish-chandra/
23. Miroslav Radenkovic,1 Ivana Lazarevic, 2 Marko Stojanovic, Tanja Jovanovic, Ethical Challenges Related to Marketing

Drugs, Trivent Publishing, 2019, https://trivent-publishing.eu/books/thebioethicsofthecrazyape/19.%20Miroslav%20Radenkoviæ.pdf

24. Jilian Clare Kohler, Martha Gabriela Martinez, Micheal Petkov and James Sale, Corruption in the Pharmaceutical Sector: Diagnosing the Challenges, Transparency International, Pharmaceutical and Healthcare Program, https://www.transparency.org.uk/sites/default/files/pdf/publications 29-06-2016-Corruption_In_The_Pharmaceutical_Sector_Web-2.pdf
25. Sophie Peresson, Programme Director, Transparency International, Pharmaceutical & Healthcare Program, Fighting Corruption in the Pharma Sector— Why We All Have A Role To Play, Transparency International, Global Health, https://ti-health.org/content/fighting-corruption-in-the-pharma-sector-why-we-all-have-a-role-to-play/
26. Mark Dunn, Governments and Companies are Failing to Address Pharmaceutical Corruption Risks, Lexis Nexis, January 1, 1970, https://bis.lexisnexis.co.uk/blog/categories/governance-risk-and-compliance/governments-and-companies-are-failing-to-address-pharmaceutical-corruption-risks
27. Mark Dunn, Sunshine Law to Shed Light on Pharma Payments, January 1, 1970, Lexis Nexis, https://bis.lexisnexis.co.uk/blog/categories/governance-risk-and-compliance/sunshine-law-to-shed-light-on-pharma-payments
28. Mark Dunn, Fine Time for Pharma: Why Due Diligence is Important, Lexis Nexis, https://bis.lexisnexis.co.uk/blog/categories/governance-risk-and-compliance/fine-time-for-pharma-why-due-diligence-is-important
29. Mark Dunn, Effective Due Diligence the Best Medicine for the Pharmaceutical Industry, Lexis Nexis, October 1, 2019, https://bis.lexisnexis.co.uk/blog/categories/governance-risk-and-compliance/effective-due-diligence-for-the-pharmaceutical-industry

30. Dr. Gopal Dabade, Shocking Corruption in Drug Companies, Deccan Herald, June 25, 2016, https://www.deccanherald.com/content/554188/shocking-corruption-drug-companies.html
31. John T. Jesseey Jr, "Pay to Prescribe": A Case for Strengthened Enforced of the FCPA in the Global Pharmaceutical Industry in 2017 and Beyond, University of Richmond, UR Scholarship Repository, 2017, https://scholarship.richmond.edu/cgi/viewcontent.cgi?article=1148&context=law-student-publications
32. Paddy Rawlinson, Immunity and Impunity: Corruption in the State-Pharma Nexus, International Journal of Crime, Justice and Social Democracy, 2017 6 (4) 86-99, https://pdfs.semanticscholar.org/6acd/9944276609c919198dc9c749910e535b7d0a.pdf
33. Fraud and the Pharmaceutical Industry, University of Wollongong, https://documents.uow.edu.au/~bmartin/dissent/documents/health/pharmfraud.html
34. Kalyan Ray, Navya P K, A Bitter Pill: The Corrupt Business of Prescription, Deccan Herald, September 4, 2022, https://www.deccanherald.com/specials/insight/a-bitter-pill-the-corrupt-business-of-prescription-1142019.html
35. Sergio Sismondo, Epstemic Corruption, the Pharmaceutical Industry, and the Body of Medical Science, Frontiers in Medical Research Metrics and Analytics, March 8, 2021, https://www.frontiersin.org/articles/10.3389/frma.2021.614013/full
36. Elizabeth A. Kitsis, MD, MBE, The Pharmaceutical Industry's Role in Defining Illness, AMA Journal of Ethics, December 2011, https://journalofethics.ama-assn.org/article/pharmaceutical-industrys-role-defining-illness/2011-12
37. J. Edgardo Campos, A Practical Approach to Combating Corruption, The Governance Brief, https://www.adb.org/sites/default/files/publication/28633/governancebrief16.pdf
38. Jeffrey Francer, Jose Zamarriego Izquierdo, Tamara Music, Kirti Narsal, Ethical Pharmaceutical Promotion and Communications Worldwide: Codes and Regulations, Philosophy Ethics and Humanities in Medicines, 9 (1):7,

ResearchGate, March 2014, https://www.researchgate.net/publication/261217849_Ethical_pharmaceutical_promotion_and_communications_worldwide_Codes_and_regulations

39. Ashwin Sapra & Biplab Lenin, India Corporate Law: Now You See it, Now You Don't! The Law on Drug Promotion and Marketing in India, Cyril Amarchand Mangaldas—Ahead of the Curve Blog, October 15, 2018, https://corporate.cyrilamarchandblogs.com/2018/10/drug-medicines-promotion-marketing-laws-india/
40. Sumi Sukanya Dutta, Pharma-Doctor Nexus Part I: Why Rules, Code for Doctors and Drugmakers Have Failed to Curb Unethical Practices, MoneyControl, September 20, 2022, https://www.moneycontrol.com/news/trends/pharma-doctor-nexus-part-i-why-rules-code-for-doctors-drugmakers-have-failed-to-curb-unethical-practices-9203591.html
41. Sumi Sukanya Dutta, Pharma-Doctor Nexus Part II: What Drives Drug Prescription in India? MoneyControl, September 21, 2022, https://www.moneycontrol.com/news/trends/health-trends/doctor-pharma-nexus-part-ii-what-drives-drug-prescription-in-india-9209041.html
42. Singh B, Jalwal P, Ruhil VK; Nexus between Medical Healthcare Professionals and Pharmaceutical Companies : Boon or bane for Patients; PharmaTutor; 2018; 6(3); 17-22; http://dx.doi.org/10.29161/PT.v6.i3.2018.17
43. Murali Neelakantan, The Doctor—Drug Company Nexus: Diagnosis and Treatment, 2022 SCC Online Blog Exp. 64. August 27, 2022, https://www.scconline.com/blog/post/2022/08/27/the-doctor-drug-company-nexus-diagnosis-and-treatment/
44. Banjot Kaur, Union Govt. Flips Stance, Tells SC, 'Don't Push for Mandatory rules to curb Pharma—Doctor Nexus', The Wire, October 4, 2022, https://thewire.in/government/government-supreme-court-pharma-doctor
45. Dr. Nida Wahid Bashir, Coordinator, Karachi Bioethics Group 2022, Karachi, Pharma—Physician Nexus is Tricky, Dawn, May 10, 2022, https://www.dawn.com/news/1688946/pharma-physician-nexus-is-tricky

46. Pramod Pathak, Dolo 650 Row Reveals Doc-Pharma Nexus, The Pioneer, November 20,2022, https:/www.dailypioneer.com/2022/columnists/dolo650-row-reveals-doc-pharma-nexus.html
47. M. Sai Gopal, Doctor—Pharma Nexus Haunts Patients, The HIndu, September 11, 2011, https://www.thehindu.com/news/cities/Hyderabad/doctorpharma-nexus-haunts-patients/article2444285.ece
48. Jyoti Shelar, Non-profits Bring to Light Doctor—Pharma Nexus, The Hindu, December 10, 2019, https://www.thehindu.com/news/cities/mumbai/non-profits-bring-to-light-doctor-pharma-nexus/article30261013.ece
49. Dr. Aneek Gupta, Unholy Nexus Between Doctors, Pharmaceutical Companies, Chemists and Pathological Clinics, SlideShare, A Scribd Company, July 28, 2009, https://www.slideshare.net/ANEEK/unholy-nexus-between-doctors-pharmaceutical-companies-chemists-and-pathological-clinics-by-aneek-gupta
50. Aleeshia Carman and Mahat Mundluru, Corruption in China, Leadership and Democracy Lab, 2016-2017 Research Archives, https://www.democracylab.uwo.ca/Archives/2016__2017_research_/pharma_in_china/corruption_in_china.html
51. John Braithwaite, Corporate Crime In the Pharmaceutical Industry, Routledge & Keagan Paul, 1984
52. Lauren Vogel, Experts Blame Feds for Pharma Corruption, CMAJ (Canadian Medical Association Journal), February 27, 2017 189 (8) E327-E328; DOI: https://doi.org/10.1503/cmaj.1095384
53. Grady Matter, The Questionable Marketing Practices of Pharmaceutical Companies, Medium, April 20, 2019, https://medium.com/@14ideas/the-questionable-marketing-practices-of-pharmaceutical-companies-5524fdbcad9b
54. Whistleblowers International, Off-Label Pharmaceutical Marketing and Unlawful Drug Promotion, https://www.whistleblowersinternational.com/types-of-fraud/pharmaceutical/off-label-marketing/

55. Lena Groeger, Big Pharma's Big Fines, Pro Publica, February 24, 2014, https://projects.propublica.org/graphics/bigpharma
56. Michelle Llamas, BCPA,Selling Side Effects: Big Pharma's Marketing Machine, DrugWatch, https://www.drugwatch.com/featured/big-pharma-marketing/
57. Sanjiv Das, How Deep runs 'The Rot' in the Pharma Sector, BioSpectrum, November 20, 2022, https://www.biospectrumindia.com/views/17/21719/how-deep-runs-the-rot-in-the-pharma-sector.html
58. RSTV: The Big Picture— Pharmaceutical Marketing Malpractices, Insights, December 10, 2019, https://www.insightsonindia.com/2019/12/10/rstv-the-big-picture-pharmaceutical-marketing-malpractices/
59. Analysis, The Biggest Ever Pharmaceutical Lawsuits, Pharmaceutical Technology, June 25, 2019, Updated August 1, 2022, https://www.pharmaceutical-technology.com/analysis/biggest-pharmaceutical-lawsuits/
60. Chris Lo, The People vs Big Pharma: Tackling the Industry's Trust Issues, Pharmaceutical Technology, August 20, 2018, https://www.pharmaceutical-technology.com/analysis/people-vs-big-pharma-tackling-industrys-trust-issues/
61. Institute for Health Policy, Drug Policy 101: Pharmaceutical Marketing Tactics, Kaiser Permanente, https://www.kpihp.org/wp-content/uploads/2020/02/drug_policy_pharmaceutical_marketing_101_FINAL.pdf
62. Pharmaceutical Fraud, CRI (Corporate Research and Investigations) Group, https://crigroup.com/pharmaceutical-industry-fraud/
63. Amanda L. Connors, Big Bad Pharma: An Ethical Analysis of Physician-directed and Consumer-directed Marketing Tactics, Albany Law Review, Volume 73, Issue 1, https://go.gale.com/ps/i.do?p=GPS&u=googlescholar&id=GALE%7CA218878426&v=2.1&it=r&sid=GPS&asid=a34f4910
64. Joan Buckley, The Need to Develop Responsible Marketing Practice in the Pharma Sector, Problems and Perspectives in Management, Research Gate, January 2004, https://

www.researchgate.net/publication/265264227_The_Need_to_Develop_Responsible_Marketing_Practice_in_the_Pharmaceutical_Sector_1

65. Beth Mole, Big Pharma Shells out $20 Billion Each Year to Schmooze Docs, $6 Billion on Drug Ads, arsTechnica, November 1, 2019, https://arstechnica.com/science/2019/01/healthcare-industry-spends-30b-on-marketing-most-of-it-goes-to-doctors/
66. Mahrukh Mohiuddin, Sabina Faiz Rashid, Mofijul Islam Shuvro, Nahitun Nahar & Syed Masud Ahmed, Qualitative Insights into Promotion of Pharmaceutical Products in Bangladesh: How Ethical are the Practices? BMC Medical Ethics, December 1, 2015, https://bmcmedethics.biomedcentral.com/articles/10.1186/s12910-015-0075-z
67. Aaron S. Kesselheim, Michelle M. Mello, David M. Studdert, Strategies and Practices in Off-Label Marketing of Pharmaceuticals: A Retrospective Analysis of Whistleblower Complaints, PLOS Medicine, April 5, 2011, https://doi.org/10.1371/journal.pmed.1000431
68. Leslie E. Sekerka, Menlo College, Deborah R. Comer, Lauren E. Benishek, Johns Hopkins Medicine, Case 1: The Inordinate Power of Big Pharma (in book: Business Ethics: How to Design and Manage Ethical Organizations, Edition 2, SAGE Publications, Editor, D. Collins), ResearchGate, January 2018, https://www.researchgate.net/publication/324603947_Case_1_The_inordinate_power_of_big_pharma
69. Meagan Parrish, Pharma's Damaged Reputation: Can the Industry Fix its Image Problem, Pharma Manufacturing, March 18, 2019, https://www.pharmamanufacturing.com/sector/small-molecule/article/11303099/pharmas-damaged-reputation
70. Sharon Yodin, Shaming Big Pharma, Yale Journal of Regulation, November 9, 2019, https://www.yalejreg.com/bulletin/shaming-big-pharma/
71. Andrew Tolve, Pharma and CSR: Why Good Deeds Are Good Business, Reuters Events Pharma, March 14, 2011, https://

www.reutersevents.com/pharma/commercial/pharma-and-csr-why-good-deeds-are-good-business

72. V. Sasirekha, Ethically Practiced Unethical Strategies in Pharma Industry— Whom to be Blamed, International Journal of Research—Granthaalayah, Vol.6 (Issue.2): February, 2018, https://oaji.net/articles/2017/1330-1520666557.pdf
73. Recent '6 Big Pharma' Frauds, BioSpectrum Asia Edition, August 6, 2013, https://www.biospectrumasia.com/analysis/25/6598/recent-6-big-pharma-frauds.html
74. Christina A. D'Souza, Upgrading the Pharmaceutical Marketing Model, ET BrandEquity, Updated on April 12, 2022, https://brandequity.economictimes.indiatimes.com/blog/upgrading-the-pharmaceutical-marketing-model/90789822
75. Arafat Mlika, Jennifer Mong, Nils Peters and Pablo Salazar, Ready for Launch: Reshaping Pharma's Strategy in the Next Normal, McKinsey & Company, December 15, 2020, https://www.mckinsey.com/industries/life-sciences/our-insights/ready-for-launch-reshaping-pharmas-strategy-in-the-next-normal
76. Scott Young, Pharmaceutical Marketing Strategies, Sermo Speaks Blog, Sermo, June 6, 2022, https://www.sermo.com/blog/sermo-speaks/pharmaceutical-marketing-strategies/
77. Utkarsh Palnitkar, Winning Strategies for Companies to Succeed in Ever-changing Pharma Market, Business Standard, April 11, 2017, https://www.business-standard.com/content/b2b-pharma/winning-strategies-for-companies-to-succeed-in-ever-changing-pharma-market-117041100541_1.html
78. Geeta Marmat, Pooja Jain, P. N. Mishra, Understanding Ethical/Unethical Behavior in Pharmaceutical Companies: A Literature Review, International Journal of Pharmaceutical and Healthcare Marketing, April 2020, ResearchGate, https://www.researchgate.net/publication/341034248 _Understanding _ethicalunethical_behavior_in_ pharmaceutical_ companies_ a_literature_review/link/61ee5dca8d338833e38f01c4/download

79. Maitri Porecha, Why Big Pharma's 100K-Strong Sales Force Is Turning Against it, The Ken, November 17, 2022, https://the-ken.com/story/why-big-pharmas-100k-strong-sales-force-is-turning-against-it/
80. Katie Thomas and Michael S. Schmidt, Glaxo Agrees to Pay $3 Billlion in Fraud Settlement, The New York Times, July 2, 2012, https://www.nytimes.com/2012/07/03/business/glaxosmithkline-agrees-to-pay-3-billion-in-fraud-settlement.html
81. Saswati Soumya Sahu, Anti Bribery Legislations' Impact on Pharmaceutical Industry, SecJure, June 28, 2022, https://www.secjure.nl/2022/06/28/anti-bribery-legislations-impact-on-pharmaceutical-industry/
82. Alexandra Wrage, Anti-Corruption Priorities for the Pharmaceutical Industry in the Age of COVID-19, forbes.com, October 22, 2020, https://www.forbes.com/sites/alexandrawrage/2020/10/22/anti-corruption-priorities-for-the-pharmaceutical-industry-in-the-age-of-covid-19/?sh=5843024d1d4d
83. Gönenc Gürkyanak Esq, Ceren Yildiz and Nazli Gürün, An Overview of Corruption Risks in the Pharmaceutical Sector, IFLR1000, February 6, 2019, https://www.iflr1000.com/NewsAndAnalysis/an-overview-of-corruption-risks-in-the-pharmaceutical-sector/Index/9221
84. Dr. Oliver Tearle, Literature: A Summary and Analysis of Washington Irving's 'Rip Van Winkle,' Interesting Literature Blog, https://interestingliterature.com/2020/05/rip-van-winkle-washington-irving-summary-analysis/?subscribe=success#subscribe-blog-blog_subscription-10
85. Disruption and the Pharma Industry: Too Big to Fail? Imperial TechForesight, March 16, 2022, https:/ imperialtechforesight.com/disruption-and-the-pharma-industry-too-big-to-fail/
86. Kevin McCaffrey, How Will the Cloud Change Pharma? MM+M, November 1, 2013, https://www.mmm-online.com/home/channel/features/how-will-the-cloud-change-pharma/

?utm_source=newsletter&utm_medium=email&utm_campaign=NWLTR_MMM_ NewsBrief_112822_GF

87. Sharon Klahr Coey, Pharma Payments to Physicians Still Deliver A Big 'Return on Investment' in Prescription Growth, Study Finds, Fierce Pharma, December 3, 2020, https://www.fiercepharma.com/marketing/more-mon

88. Aaron P. Mitchell, MD, MPH, Niti U. Trivedi, MPH, Renee L. Gennarelli, MS, Susan Chimonas, PhD, Sara M. Tabatabai, BS, Johana Goldberg, MSLIS, Luis A. Diaz Jr, MD, and Deborah Korenstein, MD, Are Financial Payments from the Pharmaceutical Industry Associated with Physician Prescribing? A Systematic Review, Ann Intern Med. 2021 Mar; 174(3): 353–361. Published online 2020 Nov 24. doi: 10.7326/M20-5665

89. V. Kumar, Transformative Marketing: The Next 20 Years, Journal of Marketing, Sage Journals, Vol 82, Issue 4, First Published online July 1, 2018, https://journals.sagepub.com/doi/10.1509/jm.82.41

90. Paul Holmes, GSK Faces Criticism Over Paxil Marketing, PRovokeMedia, January 1, 2005, https://www.provokemedia.com/latest/article/gsk-faces-criticism-over-paxil-marketing

91. Katie Thomas, J&J fined $1.2 Billion in Drug Case, The New York Times, April 11, 2012, https://www.nytimes.com/2012/04/12/business/drug-giant-is-fined-1-2-billion-in-arkansas.html

92. Aditya Kudumala, Todd Konersmann, Adam Israel, Wendell Miranda, Biopharma Digital Transformation: Gain an Edge with Leapfrog Digital Innovation, deloitte.com, December 8, 2021, https://www2.deloitte.com/us/en/insights/industry/life-sciences/biopharma-digital-transformation.html

93. Ben Hirschier, GSK, Facing Bribery Claims, Battles to Build New Sales Model, Reuters: Health News, April 17, 2014, https://www.reuters.com/article/us-gsk-corruption/gsk-facing-bribery-claims-battles-to-build-new-sales-model-idUKBREA3G16C20140417

94. Cindy A. Schipani, Junhai Liu, Haiyan Xu, Doing Business in a Connected Society: The GSK Bribery Scandal in China, Illinois Law Review, January 28, 2016, https://www.illinoislawreview.org /wp-content/ilr-content/articles/2016/1/Schipani.pdf
95. Dina El Boghdady, Pfizer Agrees to Pay $60M to Settle Foreign Bribery Case, The Washington Post, August 7, 2012, https:// www.washingtonpost.com/business/economy/pfizer-agrees-to-pay-60m-to-settle-foreign-bribery-case/2012/08/07/ a2426f5e-e0b6-11e1-8fc5-a7dcf1fc161d_story.html
96. Clark Herman, SEC Charges Lilly with Corruption Abroad: A Taste of Fines to Come, PharmExec.com, December 21, 2012, https://www.pharmexec.com/view/sec-charges-lilly-corruption-abroad-taste-fines-come
97. Richard L. Cassin, Eli Lilly Pays $29 Million in SEC Settlement, The FCPA Blog, December 20, 2012, https://fcpablog.com/ 2012/12/20/eli-lilly-pays-29-million-in-sec-settlement/
98. Linda Banks, BMS Fined $14 Million for Bribery in China, Pharmaphorum, October 6, 2015, https://pharmaphorum.com/ news/bms-fined-14-million-for-bribery-in-china/
99. Richard L. Cassin, Bristol-Myers Squibb Pays $14 Million to Resolve China FCPA Offenses, The FCPA Blog, October 5, 2015, https://fcpablog.com/2015/10/05/bristol-myers-squibb-pays-14-million-to-resolve-china-fcpa-o/
100. Press Release, SEC Charges Eli Lilly and Company with FCPA Violations, US Securities And Exchange Commission, December 20, 2012, https://www.sec.gov/news/press-release/ 2012-2012-273htm
101. Nick Fletcher, GlaxoSmithKline to Pay £297 Million Fine Over China Bribery Network, theguardian.com, September 19, 2014, https://www.theguardian.com/business/2014/sep/19/ glaxosmithkline-pays-297m-fine-china-bribery
102. Stewart Atkins, Marc Bacon, Christopher Bowe, John LaMattina, George Katzourakis, Matthew Smith, Murray Stewart, Jack Whelan, Debra Whitman, GSK's New 'Ethical' Customer Approach: Is It Delivering? A Review, Analysis and Key Lessons, Eye for Pharma

103. Scott Hensley, Staff Reporter, When Doctors Go to Class, Industry Often Foots the Bill, Wall Street Journal, December 4, 2002, https://www.wsj.com/articles/SB1038953904187251993
104. Thomas Sullivan, Supreme Court Rejection of Pfizer's Request for RICO Off Label Review: Could Open Floodgate of Cases, Policy & Medicine, May 6, 2018, https://www.policymed.com/2014/01/supreme-court-rejection-of-pfizers-request-for-rico-off-label-review-could-open-floodgate-of-cases.html
105. Beth Snyder Bulik, Senior Editor, Pharma Reputation Retains 'Halo' Even As Pandemic Media Coverage Recedes—Survey, End Points News, December 7, 2022, https://endpts.com/pharma-reputation-continues-to-shine-globally-even-as-pandemic-media-coverage-recedes-ipsos-research/
106. IFPMA, IFPMA Code of Practice 2019: Upholding Ethical Standards and Sustaining Trust, https://www.ifpma.org/resource-centre/ifpma-code-of-practice-2019/
107. Mike Cox, 8 Ways Cloud Computing Supports Pharma Marketing Operations, P360: Powered Possibilities, October 2, 2019, https://www.p360.com/swittons/cloud-computing-pharma-marketing-operations/
108. Vishal Krishna, Why Dr Reddy's Are Thinking Like A Tech Company As It Focuses on Digital Transformation? YourStory, December 4, 2019, https://yourstory.com/2019/12/aws-reinvent-dr-reddys-labs-rajdeep-ghosh-digital-transformation
109. Bodhtree Press Release, Dr. Reddy's Laboratories Reengineers Sales Process in the Cloud with Salesforce, PR.com, February 10, 2011, https://www.pr.com/press-release/296647
110. The Avenga Team, Closed Loop Marketing (CLM) in Pharma, avenga.com, December 27, 2020, https://www.avenga.com/magazine/closed-loop-marketing-clm-pharma/
111. Case Studies, Marketing Automation Leads Digital Transformation for Pfizer, Super Drive, https://www.superdrive.io/case-studies/pfizer-digital-transformation/
112. Alex Brown, Marketing Automation: How Pharma Marketers Leverage Technology for Better Results, Pharma & Life Sciences Solutions, ELSEVIER, April 26, 2022, https://www.elsme-

diakits.com/blogs/marketing-automation-how-pharma-marketers-leverage-technology-better-results

113. Matthew Herper, The Terrible Things GlaxoSmithKline Did Wrong— And The Thing It's Doing Right, forbes.com, July 2, 2012, https://www.forbes.com/sites/matthewherper/2012/07/02/the-terrible-things-glaxosmithkline-did-wrong-and-the-thing-its-doing-right/?sh=6c8a995b4607
114. Josh Meyer, Officials: Pfizer to Pay Record $2.3B Penalty, LOs Angeles Times, September 3, 2009, https://www.latimes.com/archives/la-xpm-2009-sep-03-fi-pfizer3-story.html
115. Danine Midura, 7 Benefits of Pharma CRM for Sales & Marketing, Technology Advisors, August 11, 2020, https://www.techadv.com/blog/7-benefits-pharma-crm-sales-marketing
116. Benjamin Shobert, Three Ways To Understand GSK's China Scandal, forbes.com, September 4, 2013, https://www.forbes.com/sites/benjaminshobert/2013/09/04/three-ways-to-understand-gsks-china-scandal/?sh=2d0f05de55dc
117. S. Srinivasan, Ethical Minefield: Pharma Industry's Gifts to Doctors Can be Tax Deductible, Rules Tribunal, scroll.in, January 25, 2017, https://scroll.in/pulse/827501/ethical-minefield-pharma-industrys-gifts-to-doctors-can-be-tax-deductable-rules-tribunal
118. Ethics Unwrapped, Curbing Corruption: GlaxoSmithKline in China, McCombs School of Business, University of Texas, https://ethicsunwrapped.utexas.edu/video/curbing-corruption
119. GSK China scandal. (2022, August 4). In *Wikipedia*. https://en.wikipedia.org/wiki/GSK_China_scandal
120. Ethics Unwrapped, Daraprim Price Hike, McCombs School of Business, University of Texas, https://ethicsunwrapped.utexas.edu/video/daraprim-price-hike
121. Vyera Pharmaceuticals. (2022, August 13). In *Wikipedia*. https://en.wikipedia.org/wiki/Vyera_Pharmaceuticals
122. Ethics Unwrapped, Theranos' Bad Blood, McCombs School of Business, University of Texas, https://ethicsunwrapped.utexas.edu/video/theranos-bad-blood

123. Ethics Unwrapped, OxyContin & The Opioid Epidemic, McCombs School of Business, University of Texas, https://ethicsunwrapped.utexas.edu/video/oxycontin-the-opioid-epidemic
124. Ethics Unwrapped, OxyContin: Whale Watching, McCombs School of Business, University of Texas, https://ethicsunwrapped.utexas.edu/video/whale-watching
125. Art Van Zee, MD, The Promotion and Marketing of OxyContin: Commercial Triumph, Public Health Tragedy, American Journal of Public Health, 2009 February; 99(2): 221-227, https://www.ncbi.nlm.nih.gov/pmc/articles/PMC2622774/
126. Sarah Boseley, Andrew Witty: Drug Firm Boss Out to Change His Industry, The Guardian, August 12, 2009, https://www.theguardian.com/katine/2009/aug/12/katin-glaxosmithkline-andrew-witty-pharmaceuticals
127. Shraddha Chakradhar and Cassey Ross, The History of OxyContin, Told through Unsealed Purdue Documents, STAT, December 3, 2019, https://www.statnews.com/2019/12/03/oxycontin-history-told-through-purdue-pharma-documents/
128. Fabio Turone, Italian Doctors Face Criminal Allegations Over Bribes, BMJ, June 5, 2004; 328 (7452): 1333, https://www.ncbi.nlm.nih.gov/pmc/articles/PMC420277/
129. Erika Kelton, Big Pharma's Offshore Fraud Strategy, forbes.com, September 11, 2012, https://www.forbes.com/sites/erikakelton/2012/09/11/big pharmas-offshore-fraud-strategy/?sh=702721565b0b
130. Angus Liu, With New Settlement, Novartis Has Shelled Out #1.38B for Kickbacks, Bribery and Price Fixing This Year, Fierce Pharma, July 2, 2020, https://www.fiercepharma.com/pharma/novartis-shells-out-729m-to-settle-dragged-out-u-s-kickback-charges-limits-speaker-programs
131. Gretchen Morgenson, It was his dream job. He never thought he'd be bribing doctors and wearing a wire for the feds, nbcNews, July 7, 2020, https://www.nbcnews.com/business/economy/it-was-his-dream-job-he-never-thought-he-d-n1232971

132. Eric Campbell, Ph.D, Russel L. Gruen, MD, PhD, James Mountford, MD, Lawrence G. Miller, MD, et al, Special Article:A National Survey of Physician-Industry Relationships, The New England Journal of Medicine, April 26, 2007, https://www.nejm.org/doi/full/10.1056/NEJMsa064508#t=article
133. Hannah Fresques, Dollars for Doctors: Doctors Prescribe More of a Drug If They Receive Money from a Pharma Company Tied to It, ProPublica, December 20, 2019, https://www.propublica.org/article/doctors-prescribe-more-of-a-drug-if-they-receive-money-from-a-pharma-company-tied-to-it
134. Charles Ornstein, Dollars for Doctors: Pharma Money Reaches Guideline Writers, patient Groups, Even Doctors on Twitter, ProPublica, January 17, 2017, https://www.propublica.org/article/pharma-money-reaches-guideline-writers-patient-groups-doctors-on-twitter
135. Charles Ornstein, Healthcare: After Receiving Millions Drug Company Payments, Pain Doctor Settles Federal Kickback Allegations, ProPublica, August 1, 2022, https://www.propublica.org/article/sacks-doj-dollars-for-docs-settlement
136. Ashley Wazana, Physicians and the Pharmaceutical Industry: Is A Gift Ever Just A Gift? JAMA: The Journal of the American Medical Association, January 2000, https://www.researchgate.net/publication/236307287_Physicians_ and_the_Pharmaceutical_IndustryIs_a_Gift_Ever_Just_a_Gift
137. United States General Accounting Office Report to Congressional Requesters, Prescription Drugs: OxyContin Abuse and Diversion and Efforts to Address the Problem, GAO, https://www.gao.gov/assets/gao-04-110.pdf
138. Fred Schulte, How America Got Hooked On A Deadly Drug, Fierce Healthcare, June 18, 2018, https://www.fiercehealthcare.com/hospitals-health-systems/how-america-got-hooked-a-deadly-drug
139. David Armstrong, ProPublica, Inside Purdue Pharma's Media Playbook: How It Planted the Opioid 'Anti-Story', Fierce

Pharma, November 19, 2019, https://www.fiercehealthcare.com/hospitals-health-systems/how-america-got-hooked-a-deadly-drug

140. David Armstrong, Secret Trove Reveals Bold 'Crusade' to Make OxyContin A Blockbuster, STAT, September 22, 2016, https://www.statnews.com/2016/09/22/abbott-oxycontin-crusade/
141. Lisa Schwartz, MD, Steven Woloshin, MD, MS, Medical Marketing in the United States 1997—2016, JAMA Network, January 1/8, 2019, *JAMA*. 2019;321(1):80-96. doi:10.1001/jama.2018.19320
142. Justice News Press Release, Johnson & Johnson to Pay More Than $2.2 Billion to Resolve Criminal and Civil Investigations: Allegations Include Off-Label Marketing and Kickbacks to Doctors and Pharmacists, Department of Justice Press Release, November 4, 2013, https://www.justice.gov/opa/pr/johnson-johnson-pay-more-22-billion-resolve-criminal-and-civil-investigations
143. EQS Editorial Team, Elizabeth Holmes and the Theranos Case: History of a Fraud Scandal, Integrity Line, November 21, 2022, https://www.integrityline.com/expertise/blog/elizabeth-holmes-theranos/
144. John Kotter, How to Create A Powerful Vision For Change, forbes.com, June 7, 2011, https://www.forbes.com/sites/johnkotter/2011/06/07/how-to-create-a-powerful-vision-for-change/?sh=63a90b3c51fc
145. US Department of Justice Press Release, Drug Maker Forest Pleads Guilty; To Pay More Than $313 Million to Resolve Criminal Charges And False Claims Act Allegations, September 15, 2010, https://www.justice.gov/opa/pr/drug-maker-forest-pleads-guilty-pay-more-313-million-resolve-criminal-charges-and-false
146. US Justice Department Press Release, Justice Department Announces Largest Healthcare Fraud Settlement in its History: Pfizer to Pay $2.3 Billion for Fraudulent Marketing, September

2, 2009, https://www.justice.gov/opa/pr/justice-department-announces-largest-health-care-fraud-settlement-its-history

147. Emily Willingham, Why Did Hike EpiPen Prices 400%? Because They Could, forbes.com, August 21, 2016, https://www.forbes.com/sites/emilywillingham/2016/08/21/why-did-mylan-hike-epipen-prices-400-because-they-could/?sh=4140e543280c

148. Sarah Kliff, EpiPen's 400 percent Price Hike Tells Us About A Lot About What's Wrong With American Healthcare, vox.com, August 23, 2016, https://www.forbes.com/sites/emilywillingham/2016/08/21/why-did-mylan-hike-epipen-prices-400-because-they-could/?sh=4140e543280c

149. Pat Anson, PNN Editor, GlaxoSmithKline Most Heavily Fined Drug Company, Pain News Network, November 17, 2020, https://www.painnewsnetwork.org/stories/2020/11/17/glaxosmithkline-most-heavily-fined-drug-company

150. Sherry Baker, David L. Martinson, The TARES Test: Five Principles for Ethical Persuasion, Journal of Mass Media Ethics, 16(2&3), 148-175, 2001, http://www.communicationcache.com/uploads/1/0/8/8/10887248/the_tares_test-_five_principles_for_ethical_persuasion.pdf

151. Viveka Roy Chowdhury, Will FMRAI's PIL Finally Prod Govt. to Make UCPMP Mandatory? ExpressPharma, September 1, 2022, https://www.expresspharma.in/will-fmrais-pil-finally-prod-govt-to-make-ucpmp-mandatory-editors-blog/

# Index

## A



## B



## C



## D

## K



## L



## M



## N



## O



## P

## V



## W



## Z

# ABOUT THE AUTHOR

**Subba Rao Chaganti** has a master's in business administration and over fifty-two years of pharmaceutical marketing experience covering the whole gamut and all facets of marketing, from sales management to product management to heading the total marketing activity. His experience covers domestic and international marketing and the Indian and multinational sectors.

For a few years, he taught a course on advertising and brand management at the GITAM Institute of Foreign Trade (now part of the GITAM University) at Visakhapatnam and a course on international marketing at Jawaharlal Nehru Technological University (JNTU) at Hyderabad in India as a visiting faculty.

He lives in Hyderabad, India.

Here is a list of his published books:

1. Pharmaceutical Marketing in India: Concepts, Strategy, and Cases (1990)
2. Game Plans for Post-Gatt Era: Action Agenda of the Indian Pharmaceutical Industry (1999)
3. Pharmaceutical Marketing in India, Revised Edition, (2005)
4. Compete or Forfeit: Strategies for Sustainable Competitive Edge in Pharma Product Patent Era (2007)
5. Pharmaceutical Marketing in India for Today and Tomorrow: 25th Anniversary Edition, (2018)
6. Bullseyes and Blunders: Lessons from 100 Cases in Pharmaceutical Marketing (2019)
7. Digital Pharma Marketing Playbook: Winning with the New Rules of Engagement (2020)
8. Cracking the Generics Code: Your Single-Source Success Manual for Multi-Source Products (2021)
9. Reimagine Pharma Marketing: Make it Future-Proof! (2023)
10. Brand Positioning in Pharma (2023)

www.ingramcontent.com/pod-product-compliance
Ingram Content Group UK Ltd.
Pitfield, Milton Keynes, MK11 3LW, UK
UKHW022000270726
14060UKWH00003B/603